Computational Personality Analysis

Yair Neuman

Computational Personality Analysis

Introduction, Practical Applications and Novel Directions

 Springer

Yair Neuman
Ben-Gurion University of the Negev
Beer-Sheva
Israel

ISBN 978-3-319-82587-8   ISBN 978-3-319-42460-6   (eBook)
DOI 10.1007/978-3-319-42460-6

Printed on acid-free paper

This Springer imprint is published by Springer Nature
The registered company is Springer International Publishing AG
The registered company address is: Gewerbestrasse 11, 6330 Cham, Switzerland

Acknowledgments

I would like to thank my colleagues and friends Pierre Levy, Fionn Murtagh, and Leonid Perlovsky for writing the endorsements; my research assistant Dan Assaf and my programmer Yochai Cohen for being my A-team for such a long time; my former student and present colleague Danny Livshits for involving me with fascinating professional challenges; Heim (Harvey) Hames, my friend, my new dean, and the head of the Center for the Study of Conversion and Inter-Religious Encounters for generously supporting my research on self-hate; my colleagues Lucas Lacasa and Norbert Marwan for analyzing some of my data; Rafi Malach and Newton Howard for hosting me kindly for a sabbatical leave at the Weizmann Institute of Science and the University of Oxford; and my editor Hazel Bird for professionally editing this book.

This book is dedicated to my children—Yiftach, Yaara, and Tamar—who, beyond any academic landmark, are the most significant landmarks in my life.

Contents

About the Author

Yair Neuman (b. 1968) is a full professor at the Ben-Gurion University of the Negev and currently a visiting professor at the Weizmann Institute of Science and the University of Oxford. He is the author of numerous academic papers and three books, most recently published by Cambridge University Press (2014). Professor Neuman has been involved in the cutting-edge R&D of computational tools and algorithms for studying psychological and social systems.

Chapter 1
Introduction: From *The Princess Bride* to Computational Personality

> People are very open-minded about new things—as long as
> they're exactly like the old ones.
>
> (Charles F. Kettering)

The Princess Bride (Reiner and Scheinman 1987) is a wonderful romantic fantasy directed by Rob Reiner. The film's characters, such as the Spanish fencer Inigo Montoya, who is seeking to revenge the death of his father, have become cult figures, lingering in the memories of the film's audience.

This film is perhaps a somewhat nontrivial point from which to initiate a discussion on human personality in general and on computational personality analysis in particular. However, one of the movie's scenes wonderfully explains why is it so important to understand human personality.

The scene is known as the "battle of wits," and it describes a battle of minds between the character of the "Man in Black," who is actually the hero of the film, and one of his enemies—Vizzini—who is described in the film as a genius albeit criminal dwarf of Sicilian origin.

The scene involves an encounter between Vizzini and the Man in Black. Vizzini, accompanied by his giant servant, holds captive our hero's sweetheart and threatens to kill her. Expecting an approaching clash with her savior, he admits that he cannot physically compete with the Man in Black, who is a talented fencer, but mocks the Man in Black as being intellectually inferior.

Without being qualified psychologists, we can intuitively sense that Vizzini is extremely arrogant. Describing others as inferior is a derogatory act of arrogance. In fact, when comparing his intelligence to that of Plato and Aristotle, Vizzini even describes these renowned philosophers as "morons." This is another act of devaluation and arrogance, which together form one of the characterizing features of what is known as the "narcissistic personality." Dismissing others may provide the narcissist with a sense of power. However, as we will learn later, personality is all about conflicts, and what gives you a sense of power is indicative of your innermost conflicts and soft spots.

© Springer International Publishing Switzerland 2016
Y. Neuman, *Computational Personality Analysis*,
DOI 10.1007/978-3-319-42460-6_1

Vizzini's arrogance is his Achilles' heel. This Achilles' heel is identified by the Man in Black, who challenges Vizzini to a battle of wits. This battle isn't just an entertaining competition but a duel—a battle over life and death where the winner takes all. The Man in Black pours "iocaine," a powerful odorless and tasteless poison, into two glasses of wine, which are put before the two competitors. Vizzini is ignorant of which glass contains the poison. The battle of wits begins, and will supposedly end when our genius dwarf decides where the poison is. At that point, both competitors will drink simultaneously and find out who is right and who is wrong... and dead. The battle seems to be quite simple: two goblets of wine, of which one is poisonous, and Vizzini has to choose the non-poisonous goblet. The competitors drink at the same time, and either the hero's sweetheart will be saved or she will be doomed to the company of the vicious Sicilian dwarf.

Vizzini describes this challenge as "so simple" that all he has to do is to use his superior reason in order to deduce from his knowledge of the Man in Black whether the latter is a man who would pour the poison into his own glass or into the glass of his enemy. This is a task that is typical of personality analysis. Vizzini is trying to understand his opponent's mind in order to predict the opponent's next move. From this context we gain further indications of Vizzini's narcissistic personality and of his belief in his own intellectual superiority. Let us systematically trace Vizzini's "impressive" logic, in his own words:

Argument: A clever man would put the poison into his own goblet.
Factors supporting this argument:

1. He knows that only a great fool would reach for what he was given
2. I'm not a great fool and
3. you [the Man in Black] must have known I was not a great fool.

Conclusion: "I can clearly not choose the wine in front of me."

This is an impressive expression of logical reasoning in action but not a final conclusion, as Vizzini continues to reason:

1. Iocaine comes from Australia...
2. Australia is entirely peopled with criminals
3. Criminals are used to having people not trust them
4. as you [the Man in Black] are not trusted by me.
5. Conclusion: "I can clearly not chose the wine in front of you."

Again, this is an impressive form of reasoning. However, it is not necessarily the final conclusion, as Vizzini continues:

1. "you're exceptionally strong. So, you could have put the poison in your own goblet, trusting on your strength to save you."
2. Conclusion: "I can clearly not drink the wine in front of you."
3. A qualification: "In studying, you must have learned that man is mortal." Conclusions: "[you] would have put the poison as far from yourself as possible" and "I can clearly not choose the wine in front of me."

At this point Vizzini switches between the two glasses, believing he has outwitted the Man in Black. He is amused as he believes that he has solved the riddle and tricked his opponent. His amusement is another indication of his pathological narcissism and his pathological sense of superiority.

The two men drink the wine and the Man in Black says: "You guessed wrong." Vizzini is certain that he outwitted the Man in Black and replies with a burst of manic laughter and mockery, describing the Man in Black as a fool, informing him that he has fallen victim to a classic blunder, and advising him never to confront a Sicilian when death is on the line. However, at this point Vizzini falls down dead. The Man in Black explains to Buttercup—his sweetheart—that *both* goblets were poisoned as he has spent several years building up an immunity to this kind of poison.

What have we learned from this battle of wits? First, there are minor conclusions—for instance, that it is possible to go up against a Sicilian even when death is on the line. This is, of course, not a general recommendation to challenge Sicilians but just a conclusion regarding the possibility of successfully confronting a Sicilian even when death is on the line. However, putting humor aside, we have realized the importance of *understanding human personality* and now see how this understanding must take precedence over purely logical reasoning and a game theoretic style of thinking.

The Man in Black won his life as he correctly diagnosed the *personality* of Vizzini and wisely manipulated it. An arrogant narcissistic personality cannot dismiss a challenge to its superiority. Therefore, it would have been highly reasonable of the Man in Black to expect that Vizzini would accept his challenge. In addition, it would have been reasonable to expect that Vizzini, as a narcissist trusting his intellectual superiority over others, would try to address the challenge by limiting himself to the power of pure reason and by using a kind of a rational game theory style of thinking. The idea that both goblets might contain the poison did not occur to him, although he considered that the Man in Black could be capable of absorbing the poison, a possibility immediately dismissed by another supposedly marvelous logical deduction. The idea that both goblets were poisoned was somehow pushed out of Vizzini's system of reasoning. Indeed, when playing games, we restrict ourselves to a closed set of rules. In real life, victories are sometimes achieved not by playing the game better than one's opponent but by explicitly or implicitly challenging the rules. However, Vizzini, a narcissist, was caught up in his grandiose image of his intelligence without being able to question its shortcomings. As we will learn, understanding other people is a must in matters of life and death, and this understanding has its own psycho-logic.

Understanding human personality has been important since the old days of *The Princess Bride*, but it has become more important than ever in our current individualistic and technologically oriented society. In the past, the question "*Who* are you?" was secondary to the question "*What* are you?" For the Russian nobleman, for instance, the personality of a peasant was of minor importance as the power relations between the two left no place for such particularities. This peasant—let us call him Igor—could have been an introvert or an extravert, a narcissist or obsessive, highly intelligent or uneducated, but his social role and well-structured interactions in the segregated Russian society limited the conceptualization of his individuality to a single question: "*What* are you, Igor?"

Modern democratic, complex and technologically advanced societies have brought the question "Who are you?" to the front. For example, today, marketing cannot be satisfied by the "What are you?" question (e.g. are you a man or a woman) or even by the modern bureaucratic idea of the average person—*l'homme moyen*. In today's competitive and complex free market, selling Igor a new smartphone is a challenging task that cannot be successfully accomplished only by knowing that on average men like bigger smartphones than women. In a competitive free market, brands must gain a high-level picture and specification of their potential customers.

This transformation from simplified categorizations of people to the use of high-resolution analysis is also evident in the idea of personalized medicine. Let's imagine that Igor, our Russian peasant, arrives at his local health clinic, where his physician diagnoses him as suffering from a disease of the liver resulting from heavy drinking of homemade vodka. "The good news," says Dr. Petroshkin, "is that we have a drug that can solve this problem. On average it has been shown to significantly reduce the symptoms for 80 % of patients." "Wonderful," replies Igor. "But I have read about this stuff on the internet and I was wondering: what about the rest of the patients?" "Well, this is the problem," answers our good old doctor Petroshkin. "They die in pain due to unknown complications." Igor refuses to accept this treatment. He demands that the treatment fit his own personal profile and insists that he is not treated as a general and abstract category.

Our peasant, Igor, has now become both modernized and post-modernized. He refuses to be treated through the lens of simple and general categories or through the lens of *l'homme moyen* (i.e. the average person), and demands that the medical treatment fit his own *personal and unique profile*. The demand is not for general medicine designed for the average person but for a medicine that profiles the specific individual who is the hero of the current age. For instance, instead of a general recommendation to avoid the consumption of sugar, we would like to rely on recent findings showing that our microbiome has a unique signature (Seksik and Landman 2015) and that this signature is the real information we need in order to decide whether we can enjoy unlimited consumption of ice cream or should restrict ourselves to a comparatively unexciting low-sugar option.

At this point, we are well aware of the need to understand human personality, but with this need come challenges. Personality concerns an individual's *relatively* stable patterns of thoughts, emotion and behavior. It is our naive or learned way to conceptualize in a simple manner the visible "output" of complex biological, cognitive, emotional, behavioral and cultural dimensions. However, if there are lessons to be learned from the study of complex systems, they are that the concepts we use to simplify our understanding of the world cannot be used as *explanations*, that one cannot substitute a linguistic *description* for a scientific *explanation* and that a scientific explanation is not always enough to enable the design of a *practical solution* to a certain problem. For instance, our immune system is constantly involved in making decisions about whether certain recognized biological entities in our body belong to the "enemy" (i.e. non-self) and thus should be eliminated or whether these entities belong to our "self" and should therefore be tolerated (Cohen 2000). However, describing the immune system in terms of self and non-self discrimination is just a heuristic that we use as outside observers to explain immune recognition through simple conceptual-linguistic tags. Describing the activity of the immune system through a simple binary choice (i.e. self and non-self discrimination) does not expose the complex processes involved in immune recognition and response. Similarly, understanding human personality cannot be reduced to the simplicity of a few theoretical categories per se. For instance, the current dogma in personality research is the five-factor model of personality (McCrae and Costa 2013). This model, which has been compared by its proponents to the achievements of physics in terms of its "success," describes human personality in terms of five dimensions only, such as whether one is an extravert or an introvert. Now, while modern science formalizes the processes through which certain outputs are deemed to be produced from basic building blocks, it does not exhaust the complexity of its subject matter by name calling. One would not find the "four-factor model of the human genome" or the "seven-factor model of the brain" or the "three-factor model of the bacterium," and for good reasons. Even when science identified the four letters that stand for the four components of DNA, it was clear that genetics is not about the taxonomy of species according to the four genetic letters shared by all but about something much more complex—something that, while it certainly involves a *structure* (i.e. the DNA), also, and no less importantly, involves the *dynamics* through which this structure becomes meaningful.

In addition, when trying to predict the behavior of living beings in real situations, how can we settle for just a few components? How many conceptual categories do we need in order to predict the behavior of an octopus (a quite intelligent creature)? How many features do we need to determine the behavior of a virus (no doubt an intelligent entity)? Consider how difficult it is to apply numerical weather prediction to forecast the trajectory of a storm, which, psychologists would argue, is less complex then the turbulent human personality. The bottom line is that understanding human personality for real-world applications, rather than for naive communicative or academic reasons per se, invites the perspective of complex systems for real-world and well-focused applications. Such a perspective does not exist today and is a challenge waiting to be addressed.

A further issue we should address in "profiling" a certain individual is that the digital era has enabled the formation of a wealth of data that exist on a massive scale and that invite the use of sophisticated tools for analysis. In our imaginations, in-depth understanding of personality is performed by a Freud-like expert, who interviews the individual in a one-to-one session and reaches a diagnosis through intuition guided by theorization. This scenario is impractical for massive data analysis. An intelligence agency seeking the needle of a potential terrorist in the haystack of social media and a marketing company that would like to profile potential customers in order to fit them to the most appropriate advertisement are dealing with huge numbers of individuals. The traditional methods of an "impressionist" in-depth analysis by a single individual (insightful as it may be) and the use of limited tools such as focus groups, surveys or questionnaires seem to be irrelevant to addressing the current challenges.

In sum, computational personality analysis seems to be highly important for the current era. It is a field that should be guided by a perspective that acknowledges the complexity of human personality, by huge and diverse data repositories and by automatic tools and algorithms for data analysis. It should also be motivated by a practical, pragmatic approach that respects the practically oriented process of engineering automatic systems that perform well-defined tasks. In this context, theories should serve as spaces where hypotheses and ideas are generated rather than quasi-religious spaces where *dogma* is provided to frame our thoughts and behavior.

When first introduced to the field of natural language processing, I experienced something of a cultural shock. I met one of the world's leading experts in the field, whose algorithms have produced impressive results, and I asked him about the theory guiding his work, as I expected theories to be his guiding compass. He explained that he is quite indifferent to the various psychological and sociological theories that "should" be relevant to his studies. I than realized that, as a researcher with a background in psychology, I come from a field with (too) many theories but with (almost) zero practical achievements even in allegedly simple tasks of classification. In contrast, the expert's work was shockingly poor in psychological and sociological theorization but in practice produced exceptional results. The lesson I've learned concerns the poverty of theoretical dogmatism, the importance of a pragmatically oriented approach, and the need to respect practical engineering projects, too easily dismissed by too many researchers in the social sciences.

That being said, this anecdote does not dismiss the importance of theorization. Let me explain but first apologize in advance for what might look like self-praise. The vicissitudes of life are such that, a couple of years later, the theoretically motivated algorithms that I have developed with my team, in a 1.4-billion-dollar project funded by the Intelligence Advanced Research Projects Activity, outperformed the bottom-up and theoretically free approach of the above-mentioned expert. Dogmatism, as we now understand, might exist on both sides.

The current brief manuscript doesn't aim to provide a complete review of the emerging field of automatic/computational personality analysis. Rather, it aims to introduce the reader to the field by providing the author's own limited perspective and experience in addressing this challenge in various contexts. The two main anticipated audiences of this book are, first, psychologists and social scientists who would like to learn about this exciting new field, and, second, researchers and engineers from the natural language processing, data analytics, artificial intelligence and machine learning communities who would like to understand theories of personality and how dimensions of personality can be identified automatically in textual data. For both audiences, the notion of personality as a complex system should be highly important, and several new directions for studying human personality as a complex system are proposed.

The manuscript is user-friendly and self-explanatory in the sense that the intelligent reader should be able to understand all the ideas and methodologies with almost no previous theoretical or technical knowledge. Technical issues that are necessary to understand the field are presented in such a way that even an intelligent high school student may grasp them quite easily. The ideas presented in this book are accompanied by examples and case studies. Some of the cases are fully elaborated and have been presented in published academic papers. Many of the other ideas and cases discussed or hinted at in this book have been developed in contexts where they cannot be fully exposed.

Regarding the style: I have made significant efforts to write the book in a jargon-free manner, paying tribute to the scientific simplicity and clarity that I personally appreciate and always aim to adopt in my own writings. My experience in building personality analysis systems in fields ranging from forensic analysis to customer service has made me highly sensitive to theoretical insights while at the same time fully aware of practical and evidence-based solutions. In addition, I have attempted to write a manuscript that is not only authoritative and scientifically grounded but also entertaining. Learning is a fascinating process of mental growth that inevitably involves both the pain of evolving mental schemes and the joy of acquiring new ideas and tools. There is no reason why a kitten should be joyful while learning about a new environment while an academic might suffer while studying a new topic. With this attitude in mind, let us move to the next chapter, in which the notion of personality is introduced and explained.

References

Cohen, I. R. (2000). *Tending Adam's garden: Evolving the cognitive immune self.* San Diego, CA: Academic Press.

McCrae, R. R., & Costa, P. T, Jr. (2013). Introduction to the empirical and theoretical status of the five-factor model of personality traits. In T. Widiger & P. Costa (Eds.), *Personality disorders and the five-factor model of personality* (3rd ed., pp. 15–27). Washington, DC: American Psychological Association.

Reiner, R., & Scheinman, A. (1987). *The princess bride.* USA: ACT III Communications.

Seksik, P., & Landman, C. (2015). Understanding microbiome data: A primer for clinicians. *Digestive Diseases, 33*(Suppl. 1), 11–16.

Chapter 2
Personality in a Nutshell: Understanding Who We Are

As social animals that seek to manage their relations with others, we need to understand each other. Understanding here means the interpretation of overt behavior and communicated signals but also inferences about implicit thoughts and the prediction of future moves. These mental models of others that we build in our mind are usually unconscious and include naive theories of personality. However, at a higher level of cultural development, these unconscious and naive models may turn into conscious, conceptual and even scientific theories. In addition, these mental models may start as tools for modeling others but may turn inward into models of understanding ourselves. In this context, the concept of personality may be used to describe the mental models we build about the individual's mind.

Funder (1997, pp. 1–2) defines personality as the "individual's characteristic patterns of thought, emotion, and behavior, together with the psychological mechanisms—hidden or not—behind those patterns." This theoretical definition is of course too general to enable us to resolve the enormous conceptual difficulties about the meaning of personality as discussed in the literature (Saucier 2009). In this context, we should realize that there are two different approaches to what it means to define something. The first approach, which has the flavor of what is called in philosophy "naive realism," assumes that the concept we aim to define somehow represents an a priori "real" object in the world. According to this approach, the concept of personality is a theoretical construct that aims to represent a *real* pattern or system that existed prior to any form of theorization. A different and less naive approach, which has been shown to be extremely fruitful in the history of science, suggests that our theoretical constructs and measurement processes actually *define* our object of inquiry rather than simply representing or measuring it. The ideas of imaginary numbers and potential energy are just two instances that prove the superiority of this approach over the naive one. It is also an approach that is deeply pragmatic, in the sense that it seeks to do something through the definition and measurement processes of a certain object rather than to faithfully represent a divine platonic realm that exists beyond the limited perspective of human beings. The Greek mathematicians invented the idea of irrational numbers,

© Springer International Publishing Switzerland 2016
Y. Neuman, *Computational Personality Analysis*,
DOI 10.1007/978-3-319-42460-6_2

regardless of the fact that this concept has no simple correspondence with a real-world object or its measurement. In fact, the idea of irrational numbers was found to be an extremely powerful tool *regardless* of its counterintuitive character.

The two approaches to definitions are actually caricatures used to emphasize theoretical differences. However, if we adopt the more pragmatic approach in the context of personality research, we can start with the basic and commonsense assumption that human beings should understand others (and themselves) by building integrative mental models of individuals' thoughts, emotions and behavior, in order to optimize communication, correct communication errors, predict future moves, monitor and control others and oneself, and so on. Let me explain this idea.

First, the models we build of others are "mental" in the sense that they exist in our mind. As such, they are abstract representations. These models are not necessarily conscious or sophisticated; they may just be forms of minimal representation and organization that we use in order to make sense of others and of ourselves. The models are also "mental" in a different sense. We assume that others build mental models of the world, like we do. Therefore, when building mental models of personality, we actually build mental models of mental models. This is an important point as it pertains to what is described as the *Theory of Mind*. Personality research is about how experts theorize the minds of individuals through the individual's theories of mind. A sociologist may explain the behavior of a certain individual through the concept of social class. For example, in contrast with Michael's decision to be an engineer, Patrick's decision to be a boxer may be explained by a sociologist as a behavior resulting from the fact that Patrick belongs to the working class while Michael belongs to a higher class. The explanation may be that, as the lower classes have fewer opportunities to climb the social ladder, their members are forced into careers that involve high risk and a low probability of good wages. Only a few boxers will be successful (similarly to criminals—another career choice that may be appealing to the lower classes), while being an engineer is almost certainly a safe bet. In contrast with sociologists, however, personality theorists focus on the individual's *level of analysis* and seek to explain the individual's mental models of self and others and the behaviors that result from those models. That is, the models are those of *individuals* (rather than classes, for example) and of the mental models that guide the behavior of these individuals.

For analytical reasons per se, we may consider these abstract representations/ schemes of others in terms of *thoughts, emotions* and *behavior*. By using the phrase 'analytical reasons', I mean that in practice our mental models integrate thoughts, emotions and behavior and that the artificial separation between the three concepts is done for conceptual simplicity and clarification only.

Thoughts can be described in terms of propositions that frame and guide our representations of others. Propositions are basic units of meaning that can be formalized as predicate-argument structures (Kintsch 1998). Specifically, if I think that Danny is a trusty fellow (argument), I represent a characteristic of Danny that may help me to frame, explain and predict his behavior (predicate). *Framing* is the way we organize information in a meaningful way, *explanation* involves the identification of causes and *prediction* involves the anticipation of future behavior. If I frame Danny as a "trusty" fellow, I can explain his behavior (e.g. Danny was quick to return money borrowed from me *because* he is a trusty fellow) and predict his behavior (e.g. I can give a loan to Danny as I'm almost sure that he will pay it back).

Thoughts of course are deeply intermingled with behavior. When I taste a piece of food, I can differentiate between a sweet taste and a rotten taste, and these thoughts may lead to differential behaviors. Thoughts are basically about categorization (Harnad 2005)—that is, knowing that something is X and something else is Y. Experiencing the taste of chocolate is therefore a thought/cognition/categorization deeply associated with behavior (e.g. attraction to sweet food). Food experienced as rotten, which is another form of categorization/thought, would lead to a diametrically opposed behavior, which is repulsion.

Thoughts are deeply intermingled with emotions too. The self-defeating thought "I'm a loser" is a proposition or a belief about myself that is loaded with a strong and negative emotional valence, where valence is the attractiveness (i.e. positive valence) or aversiveness (i.e. negative valence) of an object, event or situation. Thoughts, however, are more about the objects that occupy an individual's mind and their attributes and relations, while emotions are more about the valence associated with these propositions and behavior is more about the actual activities associated with thoughts and emotions (e.g. distancing oneself from what is represented as a 'disgusting' negatively loaded object).

At the most basic level, emotions are about whether the represented object, attribute or action is positive or negative (i.e. its valence) and to what extent (i.e. its arousal). Emotions are deeply associated with feedback loops and reinforcement learning. Sweetness is the result of sugar activating specific receptors on our tongue. The valence associated with this representation is in itself a representation that motivates and guides our behavior. Through reinforcement learning, sugary taste and its highly positive valence may lead to a positive feedback loop via which we might become addicted to sugary taste and as a result suffer from obesity.

In sum, this discussion has so far assumed that human beings build mental models of others' mental models and that a generic term for these models is "personality." Up to now we have been talking about naive theories of personality rather than about scientific theories, which describe mental models of individuals' relatively stable patterns of thought regarding the thoughts, emotions and behaviors of other individuals. The task of scientifically modeling these patterns is the *raison d'être* of academic personality research. The following sections aim to present some of the major models and theories of personality. It must be kept in mind that, following the pragmatic meaning of definition and measurement that I presented above, the value of personality models and theories is to be found in the outcomes of the measurement processes derived from these theories. For example, suggesting that human beings are extraverts or introverts is meaningless without a well-structured, reliable and valid procedure for measuring this personality trait and using the outcome of this procedure in a meaningful way. We may have a theoretical definition of extraversion but the next step is to have an operational definition that specifies the procedure through which me may measure this trait. Given the results of this measurement process, we may ask how helpful these results are for achieving a specific task such as improving a personalized recommender system, tracing the trajectory of a mental health problem and so on. We will keep this pragmatic approach in mind as we move on to the most popular model of personality, the five-factor model.

2.1 The Five-Factor Model of Personality

I am introducing the five-factor model of personality (FFM) , also known as the Big Five, first because it is described by its proponents as the default model of personality structure (McCrae and Costa 2013) and because it is the most popular model used in automatic personality analysis.

To understand the model, we first have to understand one of the major difficulties in personality research, which is how to identify basic personality traits. The *lexical approach* to personality research (e.g. Wiggins et al. 1988) suggests that human naive use of language encodes indispensable information about our mental models of others and that this information is the key to identifying basic personality traits.

In our daily use of language, we use numerous terms to describe others: bitter, liberal, arrogant, friendly and so on. Now, we can identify all of these descriptive terms, present them to a representative sample of subjects and ask the subjects to rate the extent to which each term describes them. In a variation of this procedure, we may ask subjects to rate the extent to which descriptive terms represent other subjects, but the basic logic of this procedure is the same. At this point, we have self-reported measurements and we may try to examine how these measurements cluster together into several chunks in order to uncover basic factors of personality. Is it possible that terms such as "self-critical" and "depressed" are clustered together? Or terms such as "friendly" and "sociable"? The most popular statistical methodology for producing such clusters of personality is *factor analysis*. It is argued that, when factor analysis is applied to the above types of questionnaires that measure self-reported descriptive terms, five "factors" or "dimensions" arise (McCrae and Costa 2013). These factors are titled "neuroticism," "extraversion," "openness to experience," "agreeableness" and "conscientiousness."

Neuroticism is about being anxious, nervous and worried and in general about being emotionally negative. The complementary dimension to neuroticism is described as "emotional stability." Extraversion is defined by sociability, assertiveness, cheerfulness and energy. It is about being energetic, assertive and forceful. The complementary aspect to extraversion is "introversion" which relates to being shy, cautious and unassertive. Openness to experience is described using terms such as "reflective," "imaginative" and "unconventional," while being closed to experience is associated with terms such as "conventional" and "rigid." Agreeableness describes an underlying tendency to be cooperative and considerate to others and is associated with empathy, trust and altruism. It is contrasted with being selfish and arrogant. Conscientiousness is about being organized, self-disciplined and ordered and it is contrasted with being disordered.

We can think about the Big Five, just like any other personality factors, as continuous dimensions (i.e. traits) or as discrete categories (i.e. personality types). In fact, thinking about the Big Five as personality types, we can almost immediately recall certain characters who are prototypes of specific personalities. For example, the American film director Woody Allen is a prototype of a neurotic personality and the famous boxer Mohammad Ali was in his heyday a prototype of an extravert. Pope Francis is an agreeable personality, the cartoon character Homer Simpson is a prototype of a disorganized personality (i.e. the complementary aspect of conscientiousness) and the famous traveler Marco Polo was probably very open to experience.

As can be seen, the FFM is a very simple and appealing model of personality that captures highly intuitive components of what we describe as personality. However, this model is imbued with difficulties, as reviewed in Neuman (2015). Let me present just a few of these difficulties, which, as a result of the model's dominance, have been somehow dismissed by mainstream psychology.

The first problem is that the FFM is based on relations between *variables* while personality lies within the *individual*. The shift between these two levels of analysis is far from trivial and has detrimental consequences for the validity and use of the model. For instance, Molenaar and Campbell (2009) have analyzed the data of 22 subjects who were measured on 90 consecutive days using multiple equivalent versions of the Big Five questionnaire. That is, each subject was asked to fill a personality questionnaire on each day. A quite shocking result of this study was that the variation between individuals didn't match the overall measurements, which means that the results found at the group level of analysis were invalid at the individual level of analysis. Think, for example, about measuring the correlation between being happy and being sad at the group level of analysis. It is fairly obvious that across individuals the correlation between these two variables will be highly negative, as 'happy' and 'sad' are antonyms and one cannot be happy and sad at the same time. However, when we drill down to the individual level of analysis and measure a specific individual across time, we may find that the impressive negative correlation has faded away. The reason is simple. Only a negligible number of subjects are consistently very happy or very sad. In the first case we may call them manic (or too happy) and in the second case we may deem them to be depressive (or too sad), and, if they abruptly shift between the two extreme poles, they suffer from manic depression disorder. Most people, however, moderately fluctuate between the poles of being happy and being sad, and therefore the impressive correlation that we found at the group level of analysis may be relevant to sociological research, or for understanding a negligible part of the population, but not for understanding the personalities of most individuals or the underlying factors of those personalities.

There are other difficulties with the FFM as well. Neuroticism is one of the most important dimensions of the Big Five. It has been found to be significantly correlated with fear, sadness and anger (Davis and Panksepp 2011). Is it possible that what we call neuroticism is mainly about feeling negative emotions or having negative thoughts? I suspect that this is the case and that, therefore, the dimension of neuroticism can be reduced to a much more basic dimension. Now we can better understand why the FFM has been so successful in academic psychology. As academic psychology mainly deals with finding differences between groups (e.g. the difference between men and women in extraversion) and finding correlations between variables (e.g. the correlation between neuroticism and suicidal behavior), and as the basic experience of feeling negative or positive emotions is correlated with so many variables, it is almost guaranteed that researchers will find statistically significant correlations between negative/positive emotions and neuroticism. However, this book deals with automatic personality analysis, which is a practical field. From this perspective the statistical significance of correlations at the group level of analysis or variations found among the items of a self-report questionnaire might be of minor importance for real-world applications. Real-world applications are mostly concerned with classification and prediction at the *individual's level of analysis*. For example, when developing a system to automatically distinguish extraverts from introverts based on voice recognition, we would like to know how successful the system is in *classifying* subjects as either extraverts or introverts.

Let us assume that we have a balanced corpus of 500 extraverts and 500 introverts who have been diagnosed as such by three expert psychologists. Let us also assume that the psychologists' judgments have high reliability, meaning that they agree between themselves on who is an extravert and who is an introvert. In our experiment, each subject is asked to read aloud three pieces of text while his voice is recorded and analyzed. Next, we try to use the features extracted from the analysis of the voice to decide whether each of the given subjects is an extravert or an introvert. In this context, we may use measures such as precision and recall. Precision asks how many of the subjects our system automatically identified as extraverts (for instance) were previously diagnosed by the psychologist as extraverts. Recall, on the other hand, asks how many of the extraverts (for instance) our system successfully identified out of the total number of extraverts in our corpus. One seldom finds psychological studies in personality in which the researchers successfully apply a classification procedure, testing their results by using measures such as precision and recall. This is a disturbing state of affairs for someone who would like to develop real-world systems based on psychological knowledge, and it explains why the majority of published studies in psychology cannot be replicated (Open Science Collaboration 2015). Let us focus our critique by using a single example.

Let us assume that the FBI is seeking to screen potential school shooters for in-depth inspection by automatically analyzing the social media materials of a massive number of subjects. In this context, the FBI is not interested in correlations between variables but in successfully identifying *individuals* who may pose a threat to the public's safety. In this case, precision is a highly important factor as it asks how many of the cases our system identified as a threat are really a danger to public safety according to some agreed criteria. In this context, correlations at the group level of analysis are of no use, or is a statistically significant regression analysis.

The problem is powerfully evident in any attempt to identify rare events, and particularly in the case of ethnic profiling. For example, for MI5, it is clear that there is a statistically significant association between religious affiliation and domestic terrorism. There is no need for a sophisticated psychological analysis. However, this association is irrelevant for the purposes of screening potential terrorists; while the probability of being a Muslim given that one is a terrorist is very high, the probability of being a terrorist given that one is a Muslim is very low. In this case, the statistically significant association is of no use unless it is somehow accompanied by a valid procedure for identifying the needle in the haystack.[1]

In sum, on the positive side, the FFM is simple, intuitively appealing and represents some psychological traits that have been known since antiquity (e.g. neuroticism and extraversion). On the negative side, it is doubtful whether, given the critiques of the FFM (e.g. Block 1995) and its over-simplicity, it is of significant relevance for real-world applications. I have used the FFM in several of my studies but consider this use as the very first step in automatic personality analysis and not as either the ideal model or the final step.

2.2 The Psychodynamic Approach

The psychodynamic approach to personality has its roots in psychoanalysis, although it has been substantially transformed since the early days of Freud and his seminal work. The basic idea of personality, as adopted by the modern psychodynamic approach, is the same as the general one presented by Funder (1997; see the opening of this chapter). However, it focuses on personality as the ways in which "we habitually try to accommodate to the exogenesis of life" (Alliance of Psychoanalytic Organizations 2006, p. 18). This is a functional approach to personality; life is imbued with anxiety and human beings have developed ways of defending themselves against anxiety. These well-structured coping strategies are actually what personality is all about. Although personality cannot be reduced to our coping styles, such styles constitute a major part of our personality. When they are used in a non-adaptive, harmful and painful way, they are called "personality disorders." For example, let's assume that Sarah suffers from social anxiety. Whenever she faces a situation in which she is expected to interact with other people, Sarah experiences a strong negative arousal accompanied by thoughts of mockery and humiliation (e.g. "They are going to laugh at me") and as a result adopts a coping style of avoidance by distancing herself from social and potentially painful interactions. Sarah would have liked to interact with other people but her coping repertoire includes only one strategy, which is total avoidance. It is quite legitimate to avoid certain social interactions, but avoiding all kinds of human

---

[1]The needle-in-the-haystack problem may seem to be unsolvable, but in fact there are creative ways to approach it successfully.

interaction is a rigid and non-adaptive strategy. As a result of using this coping strategy indiscriminately, Sarah might experience deep loneliness and pain and be categorized as suffering from a personality disorder.

Traditionally, the psychodynamic approach has been focused on personality disorders and not on healthy, normal personalities. This is a major obstacle for its application in computational personality analysis. However, this approach is theoretically grounded and presents concepts that may be used to analyze normal human beings and not only extreme cases of personality disorders. The personality disorders discussed by the psychodynamic literature may therefore be understood as the non-adaptive patterns of thoughts, emotions, relations and defensive functioning that exist among all human beings. In this context, automatically analyzing texts in order to identify defensive mechanisms and using these mechanisms to measure the distance between a specific text or person and one of the prototypes identified by the modern psychodynamic approach (Westen et al. 2012) may be of great value for various real-world applications.

Defense mechanisms can be more or less sophisticated as a function of the degree to which they distort reality. *Splitting*, for instance, is a primitive defense mechanism in which positive and negative qualities of the self and others are separated to such a degree that they may form two distinct objects: a bad object and a good object. For example, let's imagine that an infant is asking his mother for a piece of candy. When the mother refuses to give him the candy, the infant may burst into tears and experience deep rage against his mother, thinking of her as "bad." However, a couple of hours later he calms down and, when his mother embraces him, he feels secure and his mother is conceived as "good." According to the theory of Melanie Klein, this situation in which the mother is both good and bad is too complex for the infant to understand and he therefore "splits" the representation of his mother into two objects: the good mother and the bad mother. In other words, the difficulty he experiences in processing the complexity of life is resolved through the use of a simple, distorting and therefore primitive mechanism of defense. While there exists no clear supporting evidence of the developmental process described above, the idea of splitting in its general sense is of high diagnostic value. Those who describe women in terms of saints or whores, those who see the world as a battle between Satan and God, and those who oscillate between unrealistic positive and negative representations of themselves or others are all involved in a non-adaptive attempt to defend themselves from a threatening anxiety. As will be illustrated later, identifying the splitting mechanisms in texts may be of high diagnostic value in identifying potential offenders such as violent political radicals.

Splitting is an example of a primitive defense mechanism. As we mature and hopefully learn to cope with the anxieties of life in more sophisticated ways, we may adapt different mechanisms, such as humor. Humor allows us to see a painful experience from a different perspective that doesn't deny the reality of this experience but *reframes* it in a more digestible form. While humor is usually discussed in the psychodynamic literature as a sophisticated defense mechanism, I believe that it is a sophisticated defense mechanism only when it comes in a sophisticated form.

The jester, for instance, was known since early Ancient Greece as someone who made his living by making fun of other people for the amusement of his master. His humor was actually a form of ridicule, which is, as has been insightfully analyzed by Billig (2005), deeply associated with embarrassment and social control. That is, the humor used by a jester cannot be considered as a sophisticated form of defense mechanism but was rather a tool for ridiculing and socially embarrassing people in order to establish the social status of the jester's master. It was a sophisticated form of social aggression and control in the form of cultural and verbal sophistication. Along the same lines, racial jokes are not an indication of a sophisticated defense mechanism in action. In contrast, when Freud was interrogated by the Nazi Gestapo, he commented on this situation by writing with heavy irony that he heartily recommended the Gestapo to everyone. In this case, the use of humor was a clear defense mechanism used by Freud to protect himself against the anxiety caused by his encounter with the Gestapo. Humor, as we can see, may be considered to be a sophisticated defense mechanism when it deals with our own self (self-humor) in a reflective, sophisticated and sublime manner.

The Alliance of Psychoanalytic Organizations (2006) explains personality types and disorders by using several dimensions: central tension/preoccupation, affect, beliefs about self and others, and central ways of defending. The dimension of central tension/preoccupation suggests that human personality is organized around an axis or around personality themes. For example, the paranoid personality is organized around the theme of attacking or being attacked by humiliating others. Think, for example, about some of the texts produced by radical Islamists who turn their rage against the United States. A specific case, such as the Shia Hezbollah movement in Lebanon, may be of great interest, as these people consider the United States to be the "Great Satan" (a term originating in Iran) and describe Americans as a bunch of Christian crusaders who are operating aggressively and in a humiliating way against the Islamic world. Therefore, they argue, the United States is a legitimate target for an attack. Now, is this description of any psychological value? After all, these Islamists' ideology may be considered by some to be legitimate and is grounded in (albeit not necessarily justified by) the life experiences of some Muslim populations that have suffered as a result of Western imperialism.

Here, we should understand an important point (equally applicable to thinking about individuals and groups), which is that a personality theme is powerfully evident whenever *an incoherent aspect of a narrative* is identified. In the case of Hezbollah, there is strong awareness of the dilemma between attacking or being attacked by the American imperialist crusaders while there is a *total denial* of the fact that the Russians (who might also be termed "crusaders") have for many years had an imperialist hold in Lebanon, which is evident in the military base in the northern city of Latakia and in the Russians' violent actions against the Muslim population in Syria as a part of their support of the Syrian regime. In other words, the point at which we identify a psychological theme of interest is the point at which we identify a certain fixation (i.e. theme) and holes (i.e. incoherence) popping up in the narrative constituting this major theme of personality. After all, if active Islamists such as affiliates of Hezbollah strive to rebel against the American crusaders, how is it that they peacefully tolerate the existence of the Russian crusaders?

With regard to the paranoid personality, in this example, we see it expressed in the theme of attacking or being attacked, in beliefs that the self is hated and that others are potential attackers and abusers, in feelings of fear and rage, and in the use of the defense mechanism known as "projection," which entails the attribution of certain parts of one's self to others. For example, since the paranoid feels enormous aggression against others, he attempts to get rid of this aggression by attributing it to others. The paranoid is also occupied with trust issues and has a suspicious personality, as he suspects others will attack, abuse and humiliate him. However, the paranoid sometimes oscillates to the other pole: ultimate trust, which is just as unrealistic as his extreme suspicion. Think, for example, about the followers of conspiracy theories. They usually attribute malevolent intentions to others whom they claim are the hidden cause behind important events. It has been firmly established that the Al-Qaeda terror organization, under the leadership of Osama bin Laden, was responsible for the terrorist attacks of September 11. However, the zealous followers of the conspiracy theories surrounding September 11 trust neither the "official" version of the story nor any scientific refutation of their alternative theories. If these guys were authentic skeptics, they might be expected to turn their suspicion not only against the "official" version of the story but also against its non-official refutations. However, the psycho-logic of those who follow conspiracy theories is such that there is a negative correlation between the passion with which they suspect the official theories and the trust they place in various forms of alternative junk. This is a clear-cut case of paranoid thinking, which is organized around the themes of aggression and trust.

The second personality type that I would like to present is schizoid-schizotypal. In this case, the conflict concerns the desire versus the fear of closeness. The schizoid is afraid of being in love or being dependent as others might threaten his autonomy, so he defends himself via social avoidance. The schizoid seems to be detached, with an odd social appearance and peculiarities in interpersonal relations and thought processes. A prototype of this personality type is Willy Wonka from *Charlie and the Chocolate Factory*. If you have seen the film directed by Tim Burton (Grey et al. 2005), you can imagine the character of Wonka, who is socially isolated, living his life with a group of bizarre workers at his chocolate factory and avoiding any significant social interaction after a painful childhood trauma caused by his sadistic father, who was… a dentist. His appearance is odd, to say the least, and his thoughts, although creative to some extent, are clearly different from the norm, as are his human interactions, which seem to be out of touch with how adaptive people behave in society.

Here is another personality type. The antisocial "psychopathic" personality is occupied with the theme of manipulating others and avoiding being manipulated. A prototypal character of this personality is Dr. Hannibal Lecter, the cannibal psychiatrist in *The Silence of the Lambs* (Utt et al. 1991). Hannibal the cannibal enjoys manipulating the FBI agent, played by Judy Foster, with a sense of mega-lomaniac importance indicating a pathologically abnormal belief about the self.

Indeed, trying to gain omnipotent control is his major defense mechanism, as expressed in his attempt to manipulate and exploit others in cold blood with no empathy or remorse. Psychopaths don't seem to experience fear as in the game they play; they are the hunters and their victims the prey.

The narcissistic personality is one of my favorites, as, in Western culture, in which the individual has become the center of attention, you might expect to find narcissists everywhere. This personality is organized around the theme of self-esteem and the inflation versus the deflation of self-value. The narcissist holds a grandiose sense of himself that actually covers a vulnerable self. He believes that he must be perfect otherwise others will not like him. Therefore, he attempts to deal with his shaky self-esteem by idealizing some (including himself) and devaluing others, through critique and dismissive behavior. Think, for example, of the great boxer Mohammad Ali, who said that he was the greatest even before he knew it. The biblical book of Proverbs advises: "Let another man praise thee, and not thine own mouth; a stranger, and not thine own lips." From a psychological perspective, it seems that this advice aims to warn us against the danger of narcissism. Those who believe in self-superlatives and that they are the greatest, the most beautiful, the most brilliant and so on are narcissists trying to defend themselves against an inner feeling of emptiness, and, while they are praising themselves, they usually devaluate others in order to prevent any competition for the title of the greatest.

The depressive personality is organized around the theme of goodness and badness. Depressive people critically blame themselves for not being good enough and therefore experience sadness, shame and guilt. Depressive people believe that they are inherently defective and that people who really get to know them will probably reject them. Therefore, they are concerned about loss and abandonment. Groucho Marx, who humorously said that he did not want to be a member of a club that would accept people like him as members, expressed a great sense of humor— albeit humor indicating a depressive position. The depressive personality's deval-uation of the self is therefore a defense mechanism that prevents individuals from taking part in social interactions that may "prove" their inferiority. Interestingly, the Marx brothers were Jews, and this form of sophisticated, reflective and… depres-sive sense of humor seems to be a cultural mark of the early generations of Jewish immigrants to the United States.

The dependent–victimized personality is occupied with keeping and losing relationships, as such people believe they are impotent and inadequate if they are not nurtured by relationships with powerful others. Being far from such a powerful other is accompanied by the distress of separation, similar to what a baby probably feels when left alone. Therefore, the dependent personality defense mechanism involves regression to infantile behavior. Marilyn Monroe, who played the char-acter of an infantile blonde woman, might be mistaken for a dependent personality. Her *character* was that one of an infantile dependent personality, but in real life Monroe was an intelligent women who read Freud. Monroe suffered from depression, possibly resulting from the inability of her social milieu to acknowledge her real value as an intelligent individual.

The obsessive–compulsive personality is organized around the theme of control—or, more specifically, around the theme of gaining and losing control. This personality believes that anger should be controlled and that it should defend itself by isolating affect, putting emotion aside. These people are overly rational, emotionally constricted and rigid, and highly ordered and task oriented. The obsessive–compulsive personality is an excellent illustration that the same characteristics may be positive and adaptive but also in a different context self-defeating and disordering. For example, a person who is a task-oriented control freak who can put his emotions aside in order to undertake a rationally motivated action could make an excellent emergency surgeon. However, in his personal life, being a control freak might cause him to fall apart in situations where he has no full control, to feel "dead" when his emotions are put aside and to fail in close intimate relations when his rational approach is misused as a substitute for an intuitive and deep affective understanding. This is an excellent point at which to introduce the cognitive–behavioral theory of personality into the picture.

2.3   The Cognitive–Behavioral Approach

The cognitive–behavioral approach, developed by Beck et al. (1990), argues that we have genetically determined schemes that have developed to support our survival and that these schemes are information-processing frames through which we represent and interpret the world. These schemes are what actually guide our affective experience and behavior and they may be located on a scale ranging from overdeveloped to underdeveloped. For example, an obsessive personality has an overdeveloped scheme of control and an underdeveloped scheme of spontaneity. As we can see, the cognitive–behavioral approach emphasizes the importance of schemes (i.e. thoughts) that guide us and the importance of a well-balanced use of these schemes.

The personality types discussed in this approach are the same as those presented previously but the interpretation is a little different. This point can be explained through the schemes guiding the obsessive personality. To better survive and adapt, human beings should have some level of control over their lives. Cleaning your house is a form of control, as is maintaining your personal hygiene, as it allows you to avoid the harmful influence of some bacteria. Making sure your office is in order involves controlling the spatial location of objects and may help you to find them easily. Rituals are a form of control, as performing a set of actions in the same order is clearly helpful for the purposes of retrieving them from memory. However, the importance of control must be balanced by the importance of spontaneity. A mother who is trying to fully control the personal hygiene of her baby by keeping him from any contact with dirt is hurting the healthy development of his immune system and might cause the development of autoimmune diseases. The same is true for order. Order is important but, when it reaches a certain level, it blocks spontaneous and creative processes, which are at the heart of all living systems. Locating a person on the scale of order and spontaneity may help us to better understand him.

Let us present another example, which is the paranoid personality. According to Beck, all of us should maintain a healthy balance between trust and distrust. As babies we must trust our caregivers to fill our basic needs and as adults we understand that trust is a constituting aspect of social life. However, we also learn that there are some people who cannot be trusted and that trust is a highly contextual issue. What we call a paranoid personality is according to Beck a person with an overdeveloped scheme of distrust. The paranoid personality is guided by a scheme that conceives the self as vulnerable and others as vicious. The paranoid assumption is that trusting others is dangerous as it might lead to exposure to an attack. The conclusion is therefore "Don't trust." What about the narcissistic personality? According to the cognitive–behavioral approach, and following the psychodynamic approach, we should all love ourselves and think highly of ourselves. A person who has a healthy self-love may take care of himself and may consider himself highly in a way that motivates him to act in the world. On the other hand, this "superiority" should be balanced by a healthy feeling of "inferiority," putting human potential grandiosity in the right perspective. One of the greatest Hassidic Rabbis, Menahem Mendel of Kotsk, said once that a person should walk with two notes in his pocket. On the first note should be written: "This world has been created for me." This note emphasizes the narcissistic aspect of our personality and that when loving ourselves we are the center of the world. This note is to be used whenever we feel depressed. To encourage us, the note asks us to experience a large amount of self-love. The second note says: "I'm ashes and dust." It points to the inherent inferiority of human beings, who have been created from dirt and are destined to return to dirt when they die. This note is to be used whenever we experience narcissistic grandiosity that needs to be cooled down. The Rabbi of Kotsk probably grasped a very deep truth about human personality a long time before psychology became a discipline.

The cognitive–behavioral approach gives us the idea that, if we would like to understand personality, we should try to identify the *belief systems* of a person and map them onto the predefined personality types by identifying the location of the subject on each of the schemes. For example, let's assume that you would like to develop a personality-based ad-targeting system. The system should automatically identify the personality of a subject by analyzing the text of the subject's blog and filling in a predefined advertisement templates so that it will attract the attention of the specific subject. One possible advertisement could propose to aging men a new shampoo that may help them to avoid losing their hair. After identifying aging men through meta-data associated with the text, we send them the ad, which has been completed with certain keywords and phrases that may attract their attention to such a level that they will click on the ad. Different ads should be sent to different personalities. For the narcissist, who is occupied with schemes of superiority and self-grandiosity, we can use the ad: "The perfect product for the perfect man." For the paranoid personality, which is occupied with trust issues and the danger of aggression, we can use the title: "The world is a stressful place! Don't lose your hair!" The question now is how to identify the narcissist or the paranoid. As we have previously explained, the narcissist is occupied with self-love and grandiosity and therefore with issues of social superiority. Therefore, we may first try to identify in his texts words or phrases pertaining to high social status. This can be easily done by first identifying textual sources associated with the norms of the upper classes, such as *Vanity Fair* or the British *Tatler*. Using these texts, we may identify the words and phrases uniquely characterizing the upper classes and measure how far the blogger's written texts differ from the words that characterize these high-class magazines. The closer they are, the higher is the personality score on the narcissistic dimension. This is of course a very limited heuristic as one can write in a style characteristic of high-class magazines without this style having anything to do with one's personality. Therefore, a more direct procedure for measuring one's beliefs is needed. Let's say that we find the following piece of text in a blog:

> I am a brilliant person. There is no doubt that I am much more successful than most people.
> My stupid brothers were educated at a community college while I graduated from Harvard.
> I admire celebrities like me who are over and above the madding crowd.

This statement clearly may support the hypothesis that its author is ranked high on the narcissistic scale. We can process this excerpt using an automatic syntactic parser that produces the syntactic structure of a given text. Parsing the first sentence, we get the following syntactic tree:

```
(S (NP I)
   (VP am
   (NP a brilliant person)))
```

Next we can try to generate a predicate–argument structure (see the introductory section of this chapter) that allows us to identify the text's most basic units of meaning: its propositions. For example, using a simple algorithm, we can see that the noun phrase "a brilliant person" is an argument associated with the first-person pronoun "I" through the verb "am." This proposition can be represented as follows, where the arguments are located inside the square brackets and the predicate is located outside:

$$AM[I, A BRILLIANT PERSON]$$

Now, when we are trying to understand the belief system of a person, specifically his beliefs about himself, arguments associated with the first-person pronoun "I" through specific predicates may have significant diagnostic value. (At this point, we might ask how to translate the above proposition into a narcissism score, but let us leave the tricks of the trade to a later phase.)

Let us briefly summarize the psychodynamic approach and its cognitive–behavioral offspring. The psychodynamic approach is theoretically grounded. It is not a bottom-up approach like the Big Five. In addition, it is much more complex in the sense that it offers a rich repertoire of personality dimensions and their manifestation in beliefs, affect and defense mechanisms. On the other hand, and despite some attempts (e.g. Westen et al. 2012), its validity is questionable in the sense that it is not clear whether the personality dimensions and their associated aspects are really basic personality factors that are useful for understanding normal human beings specifically when taking cultural variations into account.

In sum, in this chapter, we have learned what personality is. We have taken the approach that, in order to successfully live their lives as social animals, human beings must form mental models of others' minds and of their own mind. These models aim to expose the individual's inner schemes and dynamics in a way that may explain and predict behavior. The exact nature of these schemes and dynamics is a debated issue between the various theories of personality. I have introduced two main theories of personality, the five-factor model of personality and the psycho-dynamic and cognitive–behavioral approach. There are other theories of personality. Carl Jung, who was Freud's disciple, developed his own theory of personality, which later been materialized in the Myers–Briggs type indicator (Myers 1962). The attachment theory (Bowlby 1988) identified four types of attachment that can be described as personality types. The affective neuroscience approach (Davis and Panksepp 2011) suggests that human personality can be reduced to basic mechanisms we share with other mammals, and I have presented my cognitive–biological theory of personality (Neuman 2015), arguing that all personality dimensions can be explained in terms of threat- and trust-management processes.

A person facing this rich variety of theories may respond by adopting one of two main approaches. The first approach is to choose one and only one theory or model and to dismiss the others. The second approach is relativistic: that, facing this variety of "personality" dimensions and given that there is no clear empirical test for choosing the right from the wrong, all models are equally acceptable. In contrast, the researcher or the practitioner doing automatic personality analysis may adopt a different pragmatic stance. To explain this stance, let us consider the *law of requisite variety*, originally formulated by Ashby (1958). Variety describes the various states in which a system can be. Ashby's law of requisite variety suggests that a modeling (or control) system can successfully model (and control) its domain as long as it has sufficient variety to represent that domain. Think, for example, about the immune system. The immune system has to deal with an enormous variety of pathogens, such as viruses, that rapidly genetically mutate. Now, in order to identify its enemies, the immune system has to recognize them. Immune recognition assumes some kind of modeling of the system's opponents. Therefore, our immune system has been developed in a remarkable way to produce the requisite variety of its internal "representations" (Cohen 2000).

Adopting the idea of requisite variety in the context of automatic personality analysis inevitably results in a pluralistic and pragmatic approach in which all possible dimensions of personality from whatever theoretical perspective are accepted (to a certain extent) as *features* to be used in a real-world modeling application. However, this pluralism is reduced when only relevant features are selected by our model. For example, let us assume that we would like to develop an automatic personality model that predicts who are most prone to suffering from some kind of addiction as high school students. We may ask school children to write an expository diary for a couple of months and then use analyses of those diaries to predict addiction in high school. Once we have the texts, we must identify some features that will be inserted into the model. We may use lower-level features such as patterns of letters or words (n-grams, which will be introduced in the next chapter), or more complex features such as personality dimensions measuring dependency, extraversion and so on. When using these features to predict addiction, we may find that some of them contribute nothing to the model and therefore should be removed. Other features that we haven't previously considered may surprisingly be revealed as the magic bullets of our model. The bottom line is that a pragmatic approach to automatic personality analysis optimally combines openness to different theoretical approaches and the decisiveness to choose only those features that empirically contribute to the model, given a specific task. With this approach in mind, we may move on to the next chapter, where we delve deeper into the field of computational personality analysis.

References

Alliance of Psychoanalytic Organizations. (2006). *Psychodynamic diagnostic manual*. Silver Spring, MD: Alliance of Psychoanalytic Organizations.

Ashby, W. R. (1958). Requisite variety and its implications for the control of complex systems. *Cybernetica, 1*, 83–99.

Beck, A. T., Freeman, A., et al. (1990). *Cognitive therapy of personality disorders*. New York, NY: Guilford Press.

Billig, M. (2005). *Laughter and ridicule: Towards a social critique of humour*. Thousand Oaks, CA: Sage.

Block, J. (1995). A contrarian view of the five-factor approach to personality description. *Psychological Bulletin, 117*(2), 187–215.

Bowlby, J. (1988). *A secure base: Clinical applications of attachment theory*. London, UK: Routledge.

Cohen, I. R. (2000). *Tending Adam's garden: Evolving the cognitive immune self*. San Diego, CA: Academic Press.

Davis, K. L., & Panksepp, J. (2011). The brain's emotional foundations of human personality and the affective neuroscience personality scales. *Neuroscience and Biobehavioral Reviews, 35*(9), 1946–1958.

Funder, D. C. (1997). *The personality puzzle*. New York, NY: Norton.

Grey, B., Zanuck, R. D., Siegel, M., & Burton, T. (2005). *Charlie and the chocolate factory*. USA: Village Roadshow Pictures/The Zanuck Company/Plan B Entertainment.

Harnad, S. (2005). To cognize is to categorize: Cognition is categorization. In H. Cohen & C. Lefebvre (Eds.), *Handbook of categorization in cognitive science* (pp. 20–42). Amsterdam, Netherlands: Elsevier.

Kintsch, W. (1998). Comprehension: *A paradigm for cognition*. Cambridge, UK: Cambridge University Press.

McCrae, R. R., & Costa, P. T, Jr. (2013). Introduction to the empirical and theoretical status of the five-factor model of personality traits. In T. Widiger & P. Costa (Eds.), *Personality disorders and the five-factor model of personality* (3rd ed., pp. 15–27). Washington, DC: American Psychological Association.

Molenaar, P. C., & Campbell, C. G. (2009). The new person-specific paradigm in psychology. *Current Directions in Psychological Science, 18*(2), 112–117.

Myers, I. B. (1962). *The Myers-Briggs type indicator: Manual.* Palo Alto, CA: Consulting Psychologists Press.

Neuman, Y. (2015). Personality from a cognitive–biological perspective. *Physics of Life Reviews, 11*(4), 650–686.

Open Science Collaboration. (2015). Estimating the reproducibility of psychological science. *Science, 349*(6251), aac4716.

Saucier, G. (2009). Semantic and linguistic aspects of personality. In P. J. Corr & G. Matthews (Eds.), *Cambridge handbook of personality psychology* (pp. 379–399). Cambridge, UK: Cambridge University Press.

Utt, K., Saxon, E., Bozman, R., & Demme, J. (1991). *The silence of the lambs.* USA: Orion Pictures.

Westen, D., Shedler, J., Bradley, B., & DeFife, J. A. (2012). An empirically derived taxonomy for personality diagnosis: Bridging science and practice in conceptualizing personality. *American Journal of Psychiatry, 169*, 273–284.

Wiggins, J. S., Trapnell, P., & Phillips, N. (1988). Psychometric and geometric characteristics of the revised interpersonal adjective scales (IAS-R). *Multivariate Behavioral Research, 23*(4), 517–530.

Chapter 3
Computational Personality Analysis: When the Machine Meets the Psychologist

In the previous chapter, I introduced the concept of personality and several major theories and models of personality. Traditionally, human personality has been diagnosed by a human expert who meets a person and applies his theoretical and practical knowledge in an intuitive and impressionist manner. The use of the terms "intuitive" and "impressionist" is not incidental: these terms emphasize the non-formal and non-procedural way in which some psychologists used to work and are still working today. Intuition is indispensable in understanding a single individual, but it is quite a limited approach when one has to understand a massive number of individuals. Later, and in attempts at just such an understanding, psychologists started to use self-reported questionnaires that were subject to psychometric criteria such as reliability and validity. As a result of their artificial nature, these tools have been criticized as lacking ecological validity—or relevance for real-world situations. After all, what one reports in a questionnaire and what one is actually doing are two different things.

In recent years, we have been witnessing a third wave of personality analysis that involves the use of automatic tools for the analysis of personality under the title of "computational personality analysis" (Celli et al. 2013). Computational personality analysis is usually considered to be an implementation of machine learning for the measurement and classification of personality types. Therefore, a very short introduction to machine learning is a must.

Machine learning (ML) involves the attempt to gain an abstract representation of a phenomenon from examples in order to efficiently address new future cases. More specifically, we build a model out of training examples such that it can be generalized to new cases and can successfully classify them. In "supervised ML," we label (or tag) a set of cases that are of interest and use a computer to learn an abstract model from these labeled cases. In an unsupervised form of ML, the cases are not tagged and the computer learns the model through a bottom-up process of analysis. Let us focus on supervised ML and explain how it works by using a specific example of computational personality analysis.

In the context of computational personality analysis, we may ask subjects to write an "confessional" essay representing their innermost beliefs and emotions and

© Springer International Publishing Switzerland 2016
Y. Neuman, *Computational Personality Analysis*,
DOI 10.1007/978-3-319-42460-6_3

exposing their behavior. These subjects are also assessed through a standard personality test (or an interview) that aims to score them on each of the Big Five personality factors. The result is a set of training examples in which we have (1) the subjects' essays and (2) the personality tags (e.g. neurotic) derived from the personality test. Our aim is to use various linguistic features of the essay such that, when we encounter a new, unlabeled essay, we will be able to successfully classify (or score) it into each of the five personality factors based on the model and the features.

A feature is actually a measurable property of an observed object that we would like to "learn." To address this learning challenge, we have to use a model, which is an abstract representation that describes the transformation of some input data (e.g. features of the essays) into an output (e.g. personality labels). A successful model minimizes the inevitable classification and prediction errors and can be flexibly extended to new cases. Let us examine one such potential model.

To use the essays as an input, we must first identify some relevant features that will be used as the input for the model. Let us assume that, to classify people as extraverts or introverts, we need a single feature only, which is the proportion of words in the written essay that are socially oriented. We may use a predefined lexicon in which words are categorized, such as the one prepared by Pennebaker et al. (2001), and automatically measure the proportion of social words in each essay. At this point, for each of our training examples, we have a score indicating the proportion of social words and whether the author of the essay is an extravert or not. That is, we have a continuous measurement of the feature (i.e. proportion of social words in the essay) and a binary criterion (extravert or non-extravert). At this point, we should chose a statistical model such as a binary logistic regression. This is a model in which the dependent variable is binary (e.g. extravert vs. introvert) and the independent variables may be continuous (e.g. the proportion of social words in the essay). Each model has a backbone—or fixed structure—which is determined by the way the measured variables are organized within it. For example, in a linear regression analysis with a single independent (explanatory) variable, the value of the dependent variable is described using the following structure:

$$y = ax + b$$

Here y is the dependent variable (e.g. weight) and x is the independent variable (e.g. daily calories intake). Every linear regression analysis has this general structure, which is graphically represented as a line that models a group of observations. Linear models are considered to be the simplest and the most intuitive as the amount of change in the independent variable(s) is expressed in the same amount of change in the dependent variable.

A model also includes "parameters," which are constants whose values are determined through the "learning" process in order to gain optimal performance. In the above example, the parameters are a and b. The parameter a represents the slope of the line that aims to model the linear relationship between the dependent and the explanatory/independent variables. The parameter b is the point where the line modeling our observations intersects the Y axis.

Given the decision to use a certain model such as linear regression analysis, the choice of the precise parameters is performed by an algorithm that provides the best goodness-of-fit between our model and our data by minimizing the prediction or the classification error, for instance.

Let's assume that we have applied the binary logistic regression model to our training examples. A model has been built with the appropriate parameters and now it is being applied to a new set of test examples comprising 100 extraverts and 100 introverts. The model now "guesses" the personality of each subject (i.e. extravert or introvert) by measuring the proportion of social words in the corresponding essay. To evaluate the success of the model, we may organize our results in a two-by-two table showing the actual number of extraverts and introverts by the predicted number of each class (Table 3.1).

To evaluate the model's success, we may use common measures such as accuracy, recall and precision. There are other measures of performance but these three are the basics. To recall, accuracy measures the number of cases in which the model *correctly identified* the personality tags of extraverts and introverts out of the total number of cases. As we can see, in 70 cases the model correctly classified the subjects as extraverts and in 50 cases it correctly classified the subjects as introverts. Overall, the accuracy is 0.60, or 60 %. The recall score is 70 %, because the model correctly classified 70 out of 100 extraverts. The precision score is 58 %, which reflects the number of cases identified by the model as extraverts who were actually extraverts.

Now, if you know that the proportion of extraverts in the population is approximately 50 %, this is your baseline for classifying new cases. Whenever you meet a new case, you can just toss a coin to determine whether the person is an extravert or not. In the long run this random procedure would converge to 50 % success in classification. However, the results of our model suggest that you can improve your prediction by using a single feature taken from the essays, which is the proportion of social words. Given this feature, you may increase your prediction by 8 %—from 50 to 58 %. Is such an increase in prediction good enough? There is no simple answer to such a question and the performance of the model should be judged in a rational context of costs and benefits.

In sum, then, automatic personality analysis is usually considered to be a specific instance of ML. A tagged corpus of texts is built, features are extracted, a classifier is trained and its results are evaluated. One of the early studies in the field, which illustrated this simple and basic approach, was one by Mairesse et al. (2007). These researchers analyzed a corpus of essays written by students who were also assessed

Table 3.1 Results of the classification procedure

		Predicted		Total
		Extraverts	Introverts	
Observed	Extraverts	70	30	100
	Introverts	50	50	100

using a personality questionnaire. Mairesse et al. (2007) extracted several features from the essays. In the first stage, the essays were categorized automatically into 88 word categories in the Linguistic Inquiry and Word Count (LIWC) software, such as emotion words. The authors also used 14 features from the MRC Psycholinguistic Database (Coltheart 1981), such as the familiarity of words. It was found that, using these features, various ML algorithms were able to successfully classify the subjects and that the LIWC features outperformed the MRC features for every personality trait. What we learned from this classic study is that successfully addressing the challenge of automatic personality recognition is possible and that different features and different ML algorithms may be more useful to different challenges.

This study should also raise our awareness of the importance of choosing the right features, a challenge that falls under the titles of "representation learning" and "features selection" (Guyon and Elisseeff 2003). There are various approaches to this issue. The first is to use a "fishing expedition" kind of approach and to use a "net" comprising every possible feature that may have some contribution to our model. In the context of text classification, this list of potential features may include thousands or hundreds of thousands of features. Think for instance about using each and every word in our corpus as a feature. This approach is problematic in several senses. One problem is that, when we use so many features, the "curse of dimensionality" pops up. This curse concerns various problems that become evident when we represent data in a high-dimensional space. For example, let's assume that we would like to measure the personality trait of neuroticism by measuring the frequency with which a person uses the word "sad." We can normalize this frequency to a score ranging from 0 to 100 and measure whether this feature/dimension can differentiate between neurotics and non-neurotics. However, using one feature/dimension may be too limited and therefore we may use a second dimension, which is the word "anxious." In this context, each observation (i.e. measurement of a human subject) will be represented as a point, or the end of a vector, in a two-dimensional space, as follows (Fig. 3.1).

In this case, we can see the representation of a subject who scored "75" on "sad" and "50" on "anxious." We may gather the scores of many more subjects and represent them as points in the above two-dimensional space. In this case, classifying subjects based on the two features is a little bit more tricky, as we will need to have enough subjects in our experiment and find a line that best differentiates

Fig. 3.1 Neuroticism represented as a vector in a 2D space

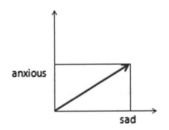

between the two groups. If we represent the subjects using three features, we should use a plane to differentiate the observations, and if we include further dimensions we will need a hyper-plane to do the job. Now, think about what happens to the volume of a space when we increase its dimensionality. It *exponentially increases* and therefore we need many more subjects for our sample and to build our model. In sum, on the one hand, gaining access to more features is important for refining our personality analysis. On the other hand and beyond a certain level, the features might turn into the curse of dimensionality.

The dimensionality issue is of great interest to researchers dealing with reinforcement learning as, when an organism faces a multidimensional stimulus, it is not always clear to which of the dimensions it is sensitive in terms of learning. For instance, if you are a devoted observer of American dramas, you may have noticed that Hollywood produces many movies that involve conflicts between couples. Observing these movies, I have asked myself many times how the poor man can be expected to learn what his spouse wants him to do, as the scenarios that lead to negative feedback from his spouse are so diverse and complicated that it seems to be impossible to represent them as a high-dimensional space from which something can be learned. Interestingly, some researchers in the field of reinforcement learning suggest that we don't process a high-dimensional stimulus as a whole but process it structurally bite by bite (e.g. Botvinick 2012; Diuk et al. 2013). Adopting the insights gained in this field may be highly relevant to ML in general and to computational personality analysis in particular.

In contrast with the fishing expedition approach to features identification as presented above, a different and more theoretically oriented approach to features selection is to use human expertise in order to identify the right features in advance, and quite a recent approach discussed under the title of "deep learning" is to let the system learn the features by itself by using representations that "are expressed in terms of other, simpler representations" (Goodfelow et al. 2016, p. 5). This approach offers an interesting solution to the problem of representation learning, but this solution has its price. As argued by Goodfelow et al. (2016, p. 17, emphasis mine):

> As of 2015, a rough rule of thumb is that a supervised deep learning algorithm will generally achieve *acceptable performance* with around *5000* labeled examples per category, and will match or exceed human performance when trained with a dataset containing at least *10 million* labeled examples.

While big datasets exist in the field of image analysis, for instance, it is difficult to imagine how a high-quality dataset of millions of examples can be simply gained in the context of computational personality analysis. There are, however, such rare cases. The work of Kosinski et al. (2013) presents an impressive dataset of Facebook users labeled according to the Big Five. The texts written by these users may be used to train a deep learning algorithm. However, this dataset is limited to the context of the Big Five and Facebook data, and it is difficult to build a validated dataset of other personality traits in other contexts.

There is another difficulty with the implementation of deep learning to personality analysis. Deep learning assumes that features can be aggregated into higher and higher representations in a straightforward (although not simple and clear) manner. However, natural language and its compositionality might challenge this assumption as it produces synergetic effects when lower-level components are clustered into bigger and bigger gestalts. Although deep neural networks have shown some remarkable results in processing textual data, these results may be improved by adhering to the synergetic nature of human language. Information synergy has been studied in the context of neuroscience and information theory and its relevance to the analysis of personality through text processing should be carefully considered, as will be discussed later. In any case, the enormous popularity of the deep learning approach should be carefully examined when looking for a solution in a concrete context of text processing. There are some alternative learning frameworks to deep learning (e.g. Tang et al. 2015; Weston et al. 2014) that have shown impressive results and should be seriously examined before designing a specific solution to a given task.

For a textual analysis, we may want to use the variety of features that appear in the scientific literature. Features can range from those that exist at the sub-word level of analysis to those that exist at the text level of analysis. For example, we may identify the n-grams—continuous sequences of n (any number of) linguistic symbols—that characterize a specific person (e.g. D A R K). The n-grams approach is also applicable to the words level of analysis. The sentence "I love to drink a shot of whisky during a rainy winter day" can be represented as a list of two-words sequence (bi-grams): "I love," "love to," "to drink" and so on. At the word level of analysis, we may also use morphological features such as suffixes. A morpheme is the most basic and meaningful grammatical unit in language. In this context, a suffix is a morpheme that is added after the word's stem, such as in the case of plurals, where the letter "s" is added to the stem (e.g. cat vs. cats). We can find the number of words that end with each kind of suffix and use it as a feature of our analysis.

The word level of analysis provides us with the category of personal pronouns (e.g. I, we, they, myself), and these have been found to be highly informative in several studies (e.g. Garrard et al. 2014). Surprising as it may sound, the use of non-content words such as pronouns and prepositions as well as other function words may be highly indicative of our personality (Koppel et al. 2011; Pennebaker 2011). At the word level of analysis, we feel much more comfortable, as words are units of meaning that usually correspond with our intuition as to what personality is. For example, we may hypothesize that extraverts use more social words and therefore that using word categories as features seems to be commonsense approach to building a personality classifier. Indeed, using word categories, such as those that appear in the LIWC, has been one of the most popular approaches in computational personality analysis.

However, while words are clearly indicative of one's personality, it has been argued that the *meaning* produced by a person is located at the proposition level of analysis and above (Kintsch 1998). Propositional representation simply involves attempts to represent knowledge using objects and predicates. Saying "cat" is just stating the existence of an object but saying "the cat is black" is a proposition that states something about the color of the cat, a proposition that can be represented in a predicate–argument form: black[cat]. If we scale up from propositions at the sentence level of analysis to the whole text, we may want to use the measure of "propositional idea density," which is defined as the ratio of expressed propositions to the total number of words in a text, a measure that has been found to be informative in diagnosing the prognosis of Alzheimer's disease through text analysis (Snowdon et al. 1996). In fact, we may use propositions as a feature that exists between the word and the sentence levels of analysis. This suggestion echoes the work of Charles Sanders Peirce, one of the greatest American thinkers, who proposed that every system of meaning can be represented by using only three categories of relations, which he named "monadic," "dyadic" and "triadic" (Peirce 1931–1966, vol. 5, p. 119).

A monadic relation is a proposition that concerns a single object (e.g. black [cat]), a dyadic relation is a proposition that involves two objects (e.g. "the cat chased the mouse," represented as chased[cat, mouse]), and a triadic relation is a proposition about three objects that *cannot be reduced* to the simple combination of two lower-order dyadic propositions. For example, the relation of "selling" necessarily assumes three objects: the seller, the buyer and the sold object. Elsewhere, I have shown (Neuman 2014) how fruitful Peirce's idea can be. As a feature we may use the proportion of the different proposition types in a text and check whether it says something about the subject who produced them.

Moving on from propositions to the sentence level of analysis, we may use grammatical structure (e.g. the syntactic complexity of a sentence) as a feature and measure. In generating a sentence, we form grammatical structures that vary in terms of their level of complexity. Military discourse may be characterized by simple grammatical structures that aim to ease communication in a cognitively loaded operational environment. However, if you read papers written by Lacanian psychoanalysts, you may be struck by their grammatical complexity, which for some of their critiques may indicate obscurity and an attempt to hide the fact that the king is naked. Syntactic complexity may be used as a feature of our ML classifier and it may be effective in recognizing military discourse, for instance. At the text level of analysis, we may use features such as the text's length, its genre (e.g. does the text have the "signature" of an academic text?) or more sophisticated measures that capture the macro-level structure of the whole text. For example, we may transform the text into a network of words and study it through various network measures, as will be illustrated later. It goes without saying that features at one level of analysis can be used to build features at higher levels of analysis: words can be grouped into categories and propositions can be used to build arguments, such as in the context of trying to identify political attitudes.

In sum, different features may result in different performances of the classifier. As we cannot know in advance which features or a combination of features will work the best, we may simply use all possible features, including those that have been theoretically selected or designed, and may let the classifier choose only those that make a significant contribution to the model. For practical applications, a delicate balance should be maintained between intuition and knowledge.

Now let's see how a computational personality classifier works in practice by recalling the first workshop on computational personality recognition (Celli et al. 2013), which took the form of a competition. The participants' main dataset was a corpus of 2468 stream-of-consciousness texts that was labeled with the personality classes of the FFM. The essays were written by students who had also been assessed through a standard inventory for their scores on each of the Big Five personality factors. The personality scores had been normalized and turned into categories (e.g. neurotic vs. non-neurotic) as the workshop focused on a classification task of the Big Five through an ML approach. The labels in the dataset were provided as categorical variables with a balanced frequency of around 50 % in each category, and each group of participants was asked to classify the subjects into their binary personality variables (e.g. extravert vs. non-extravert) using features extracted from the essays. Mohammad and Kiritchenko (2013) used several features to address this challenge. For instance, they used a "specificity lexicon" to indicate the degree to which a word used in an essay was specific or general. For example, the word "object" is general while the phrase "a double-edge shaving safety razor" is specific. They also used a lexicon of 585 refined emotion categories and, by using a specific type of classifier (i.e. support vector machine), gained the best results in the competition.

At this point, I would like to remind the reader of two important concepts that are usually missing from empirical personality research and from computational personality analysis: *context* and *time*. Let's consider context in order to describe the situational dimensions in which a measurement takes place, and time as the dynamics that characterize an observed phenomenon. When training a classifier to conduct personality recognition, researchers have used personality measures that have been collected in fixed and limited contexts and usually while ignoring the dynamics of the measured personality dimensions. Human personality, as shown in the Introduction, may have a general and relatively stable component but it is informative as long as one takes into account context (situational factors) and the flow of time. A person who is characterized as a shy introvert may show a high level of extraversion in contexts he finds to be socially welcoming and in contexts where he may express his innermost passion. A computer geek mistakenly sent to a Marines basic training session may be portrayed by his compatriots as a total introvert. However, when invited to give a talk at a conference for computer gamers, the same person may express assertiveness, energy and passion, to a level that may wash-out his previous personality stereotype.

Physicists, who supposedly study systems less complex than human personality, have remarkably incorporated such dynamic and temporal aspects into their studies, at a level that is far beyond what one may encounter in personality research. The importance of the temporal aspect can be illustrated by using the concepts of entropy versus permutation entropy One of the most fruitful scientific concepts is Shannon's information, which is explained as a measure of surprise and expressed as:

$$h(x) = -\log p(x)$$

For instance, let's assume that the chance of a rain on a winter's day is 0.99 and the chance on a summer's day is 0.01. Given rain on a summer's day, the surprise (Shannon's information) a person experiences will be significantly higher than in observing a rainy day in the winter. In this context, the concept of entropy is the surprise associated with an entire set of possible events, such as the chances of several political candidates who are running for an official position to win. Entropy is therefore the average Shannon information (Stone 2015) and it is defined as follows:

$$H(X) = \sum_{1}^{n} p(xi) \log \frac{1}{p(xi)}$$

Let's see how this measure can be applied to the analysis of personality using an example. Assume that we record the moods of two people for ten successive days. Each day we score the mood as "1" if it is generally positive and "0" if it is generally negative. Here are the time series produced for the two people:

Eli   0 0 0 0 0 0 0 1 0 0
Gadi   1 1 1 0 0 1 1 0 0 0

We can easily see that the level of uncertainty associated with Eli's mood is significantly lower than that associated with Gadi's mood, as Eli is "down" most of the time. However, consider the two following time series of mood:

A   0 0 0 0 0 1 1 1 1 1
B   0 1 0 1 0 1 0 1 0 1

These strings are equal in terms of their entropy but we can observe *two totally different temporal patterns*; while the first string represents two successive "blocks" of mood, the second string represents an oscillation between the positive and the negative moods. As Shannon's entropy fails to differentiate between these two temporal patterns of mood dynamics, we must seek measures that are more relevant to addressing this challenge, such as the method of permutation entropy (Bandt and Pompe 2002), which is able to represent sub-patterns in a string. For example, let us assume that we have a time sequence of emotions (e.g. sad) such as the one studied by Van de Leemput et al. (2014) in their study of depression. To illustrate the concept of permutation entropy, we may use seven days of measurement in which a subject reported on his level of sadness by using a Likert scale ranging from 1 (not sad) to 7 (very sad). Our time series of sadness is:

$$x = 1\,3\,2\,7\,4\,6\,1$$

The next step is to identify the "permutation order," which means the ranking in which the values appear in the series. For our case, let us select a permutation order of 3, which mean that we use a sliding window of three along the time series and examine the rank order of the values in each window. Let me explain. As our window has 3 places, there are six possible permutations (expressed as $3! = 6$), which means that the three values in a window can be organized into the following permutation (P) order, according to the ranking of their values:

$P_1 = 1\ 2\ 3$
$P_2 = 1\ 3\ 2$
$P_3 = 2\ 1\ 3$
$P_4 = 2\ 3\ 1$
$P_5 = 3\ 1\ 2$
$P_6 = 3\ 2\ 1$

Let us start moving our sliding window along the time series. The first chunk of values that we encounter is "1 3 2," which can be mapped to the permutation order 1 3 2 as this pattern expresses the ranking order of the three values. Sliding the window again one place to the right, we observe the chunk "3 2 7," which is mapped to permutation order "2 1 3" as follows:

$3 \rightarrow 2$
$2 \rightarrow 1$
$7 \rightarrow 3$

This is because the value "2" is ranked the lowest, followed by "3" and "7." If we keep moving our sliding window and counting the frequency of the different permutation possibilities, we see that the frequency of the permutation pattern P_2 is 2; that the frequency of each of the permutation patterns P_2, P_4 and P_5 is 1; and that the frequency of the other permutation patterns is 0. At this point, we may calculate the entropy of the permutation patterns and may gain a better understanding of the dynamic aspect of our time series.

Now let us go back to the starting point that initiated the discussion of the measurement of permutation entropy The features used in automatic personality analysis may be rich and diverse. Usually they have been used without reference to context and time and therefore an important aspect of understanding personality is missing. It may be interesting to use measures that express the complexity of human personality (by representing its dynamic aspect, for instance) and to use those measures as features of ML algorithms Several such ideas will be explored in the following chapters. However, before we approach the complexity of personality, let us examine simpler approaches to computational personality analysis, such as those that rely on the idea of "distributional semantics."

References

Bandt, C. & Pompe, B. (2002). Permutation entropy: A natural complexity measure for time series. *Physical Review Letters, 88*(17), 174102.

Botvinick, M. M. (2012). Hierarchical reinforcement learning and decision making. *Current Opinion in Neurobiology, 22*(6), 956–962.

Celli, F., Pianesi, F., Stillwell, D. & Kosinski, M. (2013, June). Workshop on computational personality recognition (shared task). In *Proceedings of the Workshop on Computational Personality Recognition.* Boston, MA: AAAI Press.

Coltheart, M. (1981). The MRC psycholinguistic database. *Quarterly Journal of Experimental Psychology, 33A*, 497–505.

Diuk, C., Schapiro, A., Córdova, N., Ribas-Fernandes, J., Niv, Y. & Botvinick, M. (2013). Divide and conquer: Hierarchical reinforcement learning and task decomposition in humans. In G. Baldassarre & M. Mirolli (Eds.), *Computational and robotic models of the hierarchical organization of behavior* (pp. 271–291). Berlin, Germany: Springer.

Garrard, P., Rentoumi, V., Lambert, C., & Owen, D. (2014). Linguistic biomarkers of hubris syndrome. *Cortex, 55*, 167–181.

Goodfelow, I., Bengio, Y. & Curville, A. (2016). *Deep learning.* http://www.deeplearningbook. org. Accessed April 20, 2016.

Guyon, I. & Elisseeff, A. (2003). An introduction to variable and feature selection. *Journal of Machine Learning Research, 3*, 1157–1182.

Kintsch, W. (1998). *Comprehension: A paradigm for cognition.* Cambridge, UK: Cambridge University Press.

Koppel, M., Schler, J. & Argamon, S. (2011). Authorship attribution in the wild. *Language Resources and Evaluation, 45*(1), 83–94.

Kosinski, M., Stillwell, D. & Graepel, T. (2013). Private traits and attributes are predictable from digital records of human behavior. *Proceedings of the National Academy of Sciences, 110*(15), 5802–5805.

Mairesse, F., Walker, M. A., Mehl, M. R. & Moore, R. K. (2007). Using linguistic cues for the automatic recognition of personality in conversation and text. *Journal of Artificial Intelligence Research, 30*(1), 457–500.

Mohammad, S. M. & Kiritchenko, S. (2013). Using nuances of emotion to identify personality. *arXiv preprint arXiv:1309.6352.*

Neuman, Y. (2014). *Introduction to computational cultural psychology.* Cambridge, UK: Cambridge University Press.

Peirce, C. S. (1931–1966). *The collected papers of Charles Sanders Peirce.* In C. Hartshorne, P. Weiss & A. W. Burks (Eds.) (Vol. 8). Cambridge, MA: Harvard University Press.

Pennebaker, J. W. (2011). The secret life of pronouns. *New Scientist, 211*(2828), 42–45.

Pennebaker, J. W., Francis, M. E. & Booth, R. J. (2001). *Linguistic inquiry and word count: LIWC 2001.* Mahwah, NJ: Lawrence Erlbaum Associates.

Snowdon, D. A., Kemper, S. J., Mortimer, J. A., Greiner, L. H., Wekstein, D. R. & Markesbery, W. R. (1996). Linguistic ability in early life and cognitive function and Alzheimer's disease in late life: Findings from the Nun Study. *Jama, 275*(7), 528–532.

Stone, J. V. (2015). *Information theory: A tutorial introduction.* USA: Sebtel Press.

Tang, J., Qu, M. & Mei, Q. (2015, August). Pte: Predictive text embedding through large-scale heterogeneous text networks. In *Proceedings of the 21th ACM SIGKDD International Conference on Knowledge Discovery and Data Mining* (pp. 1165–1174). New York, NY: ACM.

Van de Leemput, I. A., Wichers, M., Cramer, A. O., Borsboom, D., Tuerlinckx, F., Kuppens, P., Scheffer, M. et al. (2014). Critical slowing down as early warning for the onset and termination of depression. *Proceedings of the National Academy of Sciences, 111*(1), 87–92.

Weston, J., Chopra, S. & Bordes, A. (2014). Memory networks. *arXiv preprint arXiv:1410.3916.*

Chapter 4
Distributional Semantics and Personality: How to Find a Perpetrator in a Haystack

4.1 Vectors of Personality

I concluded the previous chapter with a call for complexity, but let us keep on introducing some simple models of computational personality analysis. A profound idea in computational linguistics is that we can represent the meaning of *words* by using vector space models of semantics (e.g. Turney and Pantel 2010). The logic behind these distributional models of semantics can be summarized using the famous expression "tell me who your friends are and I will tell you who you are." In the context of *distributional semantics*, the "friends" are words that accompany the target word that we would like to understand. According to the distributional models of semantics, representing the meaning of words is done by looking for the words that share the same context as our target word. For example, let us assume that we would like to understand the meaning of being "depressed." Understanding can have various forms. We can analytically analyze the concept of depression, as done by philosophers; we can trace its socio-cultural development, as done by historians; we can clarify its psychological sense among people; and so on. However, to understand the meaning of the word "depressed" as it is represented in textual data, we can use a corpus of texts, such as the Corpus of Contemporary American English (Davies 2009), and check which words are *collocated* with "depressed."

For example, we can search the corpus for every appearance of "depressed." When we find the word, we can use a "window" that includes the target word plus four words to the right and four to the left, and in this way we can identify the words collocated with our target—that is, the words that appear in the same lexical context as our target word. As some small words are highly frequent and attached to many other words (e.g. the, is, at), we may apply certain criteria to collocations (e.g. "mutual information") to filter out these "noisy" collocations. When searching for the collocations of "depressed," we find the following lemmas (i.e. base form) at the top of the list: anxious, feel, become, mood and patient. Following the example in

© Springer International Publishing Switzerland 2016
Y. Neuman, *Computational Personality Analysis*,
DOI 10.1007/978-3-319-42460-6_4

the previous chapter, we may represent the meaning of "depressed" as a vector residing in a high-dimensional space whose dimensions are actually the words collocated with "depressed." For example, and using a very simple and artificial example, let's represent the meaning of "depressed" by using two words only: "anxious" and "feel." In this case, our basis is {anxious, feel}. In our imagined example, the word "depressed" may appear with "anxious" five times and with "feel" eight times. Therefore, our vector is {5, 8}. A two-dimensional representation of "depressed" then looks like Fig. 4.1.

The meaning of "depressed" is here represented as a vector defined by a two-dimensional space. This is, of course, an oversimplified example. In practice, the dimensional space in which meaning is represented may be very high, and sophisticated techniques are used to build this space, to reduce it to lower and workable dimensionality and to use it for various tasks.

The most important aspect of vector space models of semantics for personality analysis is that we can use them to measure similarity between words and between texts. For instance, we can measure the similarity between "depressed" and "lonely" as the cosine between the two vectors representing the words, as shown in Fig. 4.2.

This logic can be extended to measure the similarity between vectors of texts. For computational personality analysis, this is great news as we can measure/recognize the personality that is expressed in a text by measuring the distance between the words constituting the text and a vector of words representing a given personality type. This is precisely the logic that guided my vectorial semantics approach to personality recognition (Neuman and Cohen 2014) and it can

Fig. 4.1 A 2D representation of "depressed"

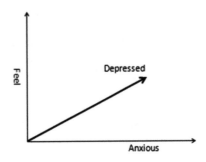

Fig. 4.2 The distance between the vector of "lonely" and "depressed"

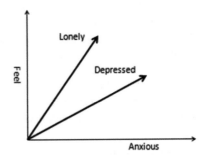

be summarized by the following steps: (1) based on theoretical and/or empirical knowledge, identify a set of words that represent a psychological trait; (2) represent this set as a vector; (3) choose a text you would like to assess and represent its words as a vector; and (4) measure the similarity between the two vectors by using a vector space. The similarity score should be indicative of the degree to which the personality trait is represented in the text.

How did my colleagues and I test this approach to personality analysis? First, we chose to test it in the context of the Big Five and their association with personality disorders. We used a meta-analysis of studies that examined the relation between personality disorders and the Big Five (see Saulsman and Page 2004). Table 4.1 presents the number of people who would score high on one of three major personality dimensions and the number who would score low on the dimension, for 100 people with a particular personality disorder.

For example, in Table 4.1, we can see that out of 100 paranoids we should expect to find approximately 64 who would score high on the neuroticism dimension and 36 who would score now on neuroticism. This result means that a comparatively high proportion of paranoids should be expected among those who are classified as neurotics. This is, of course, a rather trivial conclusion, as neuroticism is almost by definition a personality trait that is associated with personality disorders.

Indeed, the meta-analysis shows that most personality disorders have positive associations with neuroticism and negative associations with agreeableness. Therefore, my colleagues and I hypothesized that, if our vectorial semantics approach to personality assessment can validly measure the degree to which a personality dimension is evident in a text, then the relation between our personality dimension scores (as measured by the similarity of essays to personality disorder vectors) and the five factors should reproduce the above patterns.

To test this hypothesis, we used Millon et al.'s (2012, p. 4) theorization of personality disorders and extracted the first main adjectives these authors used to define personality dimensions. We used these adjectives as word vectors to represent the following personality dimensions, as follows:

Table 4.1 Personality disorders and neuroticism, extraversion and agreeableness

Personality disorder	Neuroticism		Extraversion		Agreeableness	
	High	Low	High	Low	High	Low
Paranoid	64	36			33	67
Schizoid			39	61		
Schizotypal	68	32	36	64	40	60
Histrionic			71	29	33	67
Narcissistic			60	40	37	63
Avoidant	75	25	28	72		
Dependent	71	29				

1. Schizoid: indifferent, apathetic, remote, solitary
2. Depressive: Sad, depressed, hopeless, gloomy, fatalistic
3. Avoidant: shy, reflective, embarrassed, anxious
4. Dependent: helpless, incapable, passive, immature
5. Histrionic: dramatic, seductive, shallow, hyperactive, vain
6. Narcissistic: selfish, arrogant, grandiose, indifferent
7. Compulsive: restrained, conscientious, respectful, rigid
8. Paranoid: cautious, defensive, distrustful, suspicious
9. Schizotypal: eccentric, alien, bizarre, absent.

Next, we used the essays corpus mentioned before, processed each essay and measured its similarity to each of the personality vectors (the full technical details can be found in Neuman and Cohen 2014). Overall, the similarity scores that our methodology produced replicated the general patterns presented in the above-mentioned meta-analysis. For example, we found that neurotic subjects scored significantly higher on each of the personality scores and that agreeable subjects scored significantly lower on most of the personality dimension scores, exactly as expected.

For a closer analysis of the results, let's return to Table 4.1. We can see that the proportion of paranoid, schizotypal, avoidant and dependent personalities was significantly higher among the neurotic subjects. As described above, we calculated the similarity scores of each essay to the vectors representing these personality dimensions and compared the similarity scores of neurotics versus non-neurotic subjects by using the Mann–Whitney test. Table 4.2 presents the mean rank scores for the two groups.

We can see that the neurotic subjects scored *significantly higher* on each of the personality scores. Again, these results provide some empirical support for the validity of the vectorial semantics approach to personality analysis. However, how effective are the similarity scores as features of an ML classifier that aims to recognize the Big Five personality dimensions?

To answer this question, we formed vectors for each of the Big Five personality dimensions and used all of the vectors' similarity scores as features of an ML classifier that aimed to recognize the personality types of the subjects. We used the scores showing the similarity of each essay to the personality dimensions and the Big Five vectors as features that were fed into a tree-based classification model (i.e. the classification and regression tree model). The results are presented in Table 4.3

Table 4.2 Mean Rank of personality similarity scores across neurotic and non-neurotic subjects

	Neurotic	Non-neurotic
Paranoid	1292	1177
Schizotypal	1263	1207
Avoidant	1308	1161
Dependent	1303	1166

Table 4.3 The macro-averaged F1 scores gained by MC (Mohammad and Kirichinco 2013) and NC (Neuman and Cohen 2014) for the classification task

	Ext.	Neu.	Agr.	Con.	Ope.
MK	56.28	58.25	54.20	56.56	60.61
NC	55.10	57.64	53.20	54.61	63.39

Ext. extraversion; *Neu.* neuroticism; *Agr.* agreeableness; *Con.* conscientiousness; *Ope.* openness

in terms of the macro-averaged F1 score, a measure that takes into account the precision and recall of the classifier. For a comparative analysis, the table shows the best results gained by Mohammad and Kiritchenko (2013) at the computational personality workshop (abbreviated as MK) and compares them to ours (abbreviated as NC).

The difference between the scores found by MK and NC is not statistically significant, which means that, by simply using the personality vectors and only one and quite limited classifier, my colleagues and I were able to gain results that did not significantly differ from those gained using state-of-the-art results. The modest conclusion that can be gained from these results is that the distributional semantics approach to personality analysis may work quite well. However, this approach is limited, like all other models of distributional semantics. One problem with these models is that distributional vectors often do not capture the differences in antonyms since those often have similar contexts and therefore more sophisticated approaches are needed (Socher et al. 2013). The models are also usually limited to the word level of analysis, and when we want to scale up we are in trouble. I will try to deal with these difficulties later, but let us first see how far we can push this simple approach to computational personality. The next case, that I would like to present concerns the lyrics of music genres.

4.2 Rock-and-Roll Personality

The idea that the words used by people are indicative of their personality is the bread and butter of computational personality. This idea should be taken with a pinch of salt as the language people use is not a simple representation of their minds. Human language can be used to deceive, to comply to social norms or just to express common patterns of communication such as in the case of politeness. However, notwithstanding all these accompanying difficulties, the idea that language is a gateway to personality is a necessary working assumption of computational personality, at least the part of it that deals with the analysis of texts. How far can we push this assumption? Can we argue that texts in general have some kind of personality? For example, can we argue that the lyrics of a song represent a personality? And how about the lyrics used by specific music genres? Do they have a distinct personality too? Do lyrics used by punk songs express a different personality from those used in pop? My colleagues and I have studied precisely this topic (Neuman et al. 2015).

The logic underlying this study was that there is some evidence that preference for different music genres is associated with different personalities. For example, Langmeyer et al. (2012) found that extraverts prefer energetic types of music and that individuals who are open to experience (i.e. the openness dimension of the Big Five) prefer complex music. If you think about it, this makes a lot of sense. If you are an extravert who is full of energy, you will probably prefer a music style that adheres to your temper. However, in our study, we were interested in the *lyrics* of various *music genres*, which is a much more delicate issue.

Our first study asked whether and how the words used by different personality types are similar to the words used by different music genres. For instance, we hypothesized that extraverts, in comparison with non-extraverts, will use in their essays words that are similar to the words used by energetic music, such as rock-and-roll.

To test this hypothesis and others, we used a dataset of lyrics from many songs. The songs were tagged according to their music genre, and the dataset comprised the words and the genres characterizing 17,495 songs. We also used the essay corpus mentioned before, which includes essays written by students and the personality dimensions characterizing each student. Next, we characterized each of the songs' genres by using the tf–idf (term frequency–inverse document frequency) measure. This measure aims to represent how important a word is in a given corpus. Using this measure, we ranked the words used in each genre according to their level of importance and chose the top 100 words for each genre. At this point we had vectors of words characterizing each of the music genres. The next step was to turn each of the students' essays into a vector of words and to compare the essay vectors with the music genre vectors to find similarities and differences.

The results showed that different personalities (as characterized by their essays) tended to be similar to different genres of lyrics. For example, and as hypothesized, the words used by extravert subjects were much more similar than those used by non-extraverts to the words characterizing energetic music genres such as rock-and-roll. Along the same line, it was found that the words used by neurotic subjects were more similar than those of non-neurotic subjects to the words characterizing reflective music such as jazz.

In sum, and metaphorically speaking, we found that different personalities play different music; while extraverts talk rock-and-roll, neurotic people talk the language of jazz. Moreover, we tried to classify the subjects into their personality types by using their essays' similarity to each of the music genres, and proved that one's personality can be successfully predicted based only on one's essay's similarity to the lyrics of music genres!

In fact, we found that there were only three major music genres relevant to this prediction: energetic (e.g. rock-and-roll, hip-hop), rebellious (e.g. heavy metal) and reflective (e.g. jazz). This is an interesting finding that makes a lot of sense. Music, both sound and lyrics, touches the chords of our soul, and therefore the songs we feel to be emotionally moving are indicative of who we are, probably much more than laboratory experiments or sterile personality questionnaires. If you meet a guy who tells you that the song he currently likes the most is "Feel Right" by Mark Ronson (featuring Mystikal),[1] you may hypothesize that this guy has some attraction to energetic and rebellious music, as this energetic funk song includes words such as "fuck," "shit," "bitch," "rapping," "slapping" and "cock." In contrast, you may feel this song to be quite repulsive; your own favorite song may be "Quiet Nights," specifically as performed by the wonderful jazz singer Diana Krall. This song includes words such as "nights," "stars" and "silence." In this case, we may hypothesize that, as you like reflective music, you are an introvert more than an extravert.

Taking the all above-mentioned ideas and results a step further, my colleagues and I asked whether it is possible to classify songs into their various music genres by measuring the similarity of their lyrics to the personality vectors we used in Neuman and Cohen (2014). We used the k-nearest-neighbors algorithm (KNN) for the classification procedure and each time asked whether a song belongs to a certain music genre by using our personality vectors. In other words, we asked the classifier to make a binary decision about whether a song belongs to a certain music genre or not, based on the similarity of its lyrics to each of several personality vectors. Surprisingly, we were able to successfully classify the songs based on their underlying personality, a finding that provides additional evidence of the effectiveness of using personality vectors for various computational tasks. Further evidence appears in the next section.

4.3   Finding a Murderer in a Haystack

On April 16, 2007, an American student by the name of Seung-Hui Cho entered Virginia Tech, where he was studying toward a degree. He was armed with two handguns that he had legally purchased. Cho murdered 32 people and wounded many others in a shooting incident known to be the deadliest solo attack in the United States. He didn't get out of the incident alive, shooting himself before the police could reach him. Many other violent incidents have taken place in the United States, and school shooters, despite their very low prevalence, have a terrorizing influence on the public's morale: in our perceptions, schools should be safe places, which, similarly to places of worship, should not be contaminated by human evil or violence. In reality, and paradoxically because of their sacred status, schools seem to attract murderers, who act to get even with those they consider to have inflicted harm on them.

Is there a unique psychological profile of school shooters? Interestingly, one cannot find a biometric signature or "fingerprint" of a shooter's personality. However, narcissistic personality disorder has been considered to be a central theme

---

[1]See the wonderful clip of the song at https://www.youtube.com/watch?v=ognnZ3r2qyQ.

of civilian mass murderers in general and school shooters in particular (Knoll 2010). The reason is that these shooters seem to be motivated by vengeful intentions and revenge, which have been described in the psychodynamic literature as responses to narcissistic injury. Simply stated, the narcissist is characterized by a fragile self that is hidden under a facade of grandiosity. When the narcissist is humiliated, or believes that he has been humiliated, he experiences this humiliation as an attack that might annihilate him.

During our life we might be involved in derogatory interactions. A high school student may suffer from derogatory interactions with his class mates, a young engineer may suffer derogatory comments from her chauvinistic boss, and a young scientist may be the target of derogatory criticism launched at him by his professor. No one really enjoys derogation or humiliation, but a mentally healthy person has the ability to 'digest' this painful experience without breaking into pieces. After all, behaving in a derogatory manner is indicative of a weak spot in the offender rather than of any trait in the victim. However, most people don't adopt this mature and healthy approach, and respond to derogatory behavior with anger. Stephen King grasped the relation between humiliation and revenge remarkably well in his famous novel *Carrie* (King 1974/2011), in which a young high school student with supernatural powers gets even with the school students who have humiliated her. Those who believe that school shooters suffer from a kind of malignant narcissism may believe that murderers such as Cho have the same profile as the character of Carrie; they may further believe that, when some people experience humiliation that they cannot endure, they seek revenge in order to restore their lost honor. However, while the hypothesis that school shooters suffer from malignant narcissism is theoretically grounded, there is no clear empirical evidence to support it. In addition, clinical diagnosis may seem to be interesting but for practical reasons it might be meaningless. For instance, let's assume that we know beyond any reasonable doubt that school shooters are narcissists. So what? Can we use this diagnosis to screen for potential school shooters? After all, many people experience "narcissistic injury" and have violent fantasies of getting even with those who have humiliated them. However, the overwhelming majority of people who experience such feelings and fantasies do not turn their fantasies into bloody deeds.

In a paper published in *Frontiers in Psychiatry* (Neuman et al. 2015) and covered by *Forbes* magazine, my colleagues and I applied our vectorial semantics approach to personality analysis in order to see whether it is possible to "profile" school shooters and apply this profile to screen potential shooters. To address this challenge, we chose six texts written by school shooters, including the famous "manifesto" of Seung-Hui Cho. In addition, and as a benchmark for comparison, we used a corpus of 6056 blogs written by males aged 15 to 25. We measured the semantic similarity between each of the texts and word vectors representing four personality disorder traits: paranoid personality disorder, narcissistic personality disorder, schizotypal personality disorder and depressivity. In addition to these personality vectors, we measured the texts' similarity to nine additional word vectors that covered the theme of revenge:

1. Hopeless: hopeless, desperate
2. Lonely: lonely, lonesome
3. Helpless: helpless, defenseless
4. Pain: pain, misery, agony
5. Revengeful: revengeful, vengeful, vindictive
6. Chaotic: chaotic, disordered
7. Unsafe: unsafe, insecure
8. Abandoned: abandoned, deserted
9. Humiliated: humiliated, shamed.

To make a long story short, we processed the texts of the shooters and the bloggers, measured the similarity of each text to our personality vectors and compared the scores of the shooters to those of the comparison group. It was found that the school shooters' texts scored higher on the vectors representing revenge, narcissistic personality disorder and humiliation. These results support the hypothesis that shooters suffer from malignant narcissism but they also provide us with a more complex narrative that may explain the formation of the shooter's mind.

According to this narrative, the shooter is a narcissist characterized by a fragile self. He experiences some kind of humiliation that shatters his world and as a result generates vengeful intentions that turn into a bloody deed. Now, even if this narrative is empirically grounded, how helpful can it be? To answer this question we used our personality features in three classifiers: a binary logistic regression analysis, a tree classification with a 10-fold cross-validation procedure, and the KNN with 10-fold cross-validation. For each model, we predicted the probability that a text was written by a shooter and, using this probability, ranked the texts in descending order. Table 4.4 presents the results of our analysis by showing the ranking of each shooter. The far-right column presents the mean of the three ranks.

We gain the best results by averaging the ranks of the texts' probabilities. We can see that all of the shooters' texts were ranked among the top 210 cases. Now let us assume that you are an FBI agent who somehow, and legally of course, obtains a corpus of texts published via social media. Your boss asks you to fish out of this

Table 4.4 Results of the ranking procedure

Name of shooter	BLR	Tree	KNN	Mean of ranks
Cho	1	69	1	1
Pekka	3	1	64	209
De Oliv.	5	2	184	210
Kinkel	22	79	47	19
Luke	118	119	161	56
Lepine	227	228	762	161

BLR binary logistic regression analysis; *Tree* tree classification with the chi-square automatic interaction detector technique and 10-fold cross-validation; *KNN* k-nearest neighbors analysis with 10-fold cross-validation ocean a bunch of potential shooters for a second inspection. When you check your pile, you find that it includes thousands of texts! However, if you use our automatic ranking methodology, you will find your cases among the top 210 ranked texts, which is approximately 3 % of the original corpus. This is a huge reduction in your workload. This methodology is not a magic bullet for identifying shooters in advance (indeed, if such a magic bullet exists, I'm not familiar with it), but this is a powerful archetype of a tool that may be used by law-enforcement agencies as a pragmatic solution to reduce the size of the haystack in which the needle hides.

Criticisms of this methodology were mostly ideological and appeared to focus on displaying the reviewers' own smartness at the expense of considering the methodology as a pragmatic solution. One criticism argued that, in practice, agents don't have to deal with thousands of texts but rather with hundreds of thousands of texts or even more per day. This criticism assumes that a single person is screening for shooters using a single methodology. However, what if we distribute the labor by providing each local FBI office in the United States with access to the records of its community? This sounds better, but it could be even better if we enhanced our tools with linked information from various sources. For instance, if you are a high school student with a criminal record and a history of violent behavior, sitting in his garage and producing nasty texts on social media in which you are threatening to hurt others, the FBI could receive a warning alert. If you purchase weapons and your purchase can be traced, the warning alert should turn red, and so on. The more information we gather, the better we can reduce the size of the haystack in which the potential school shooter is hiding. Therefore, while complete solutions cannot be found, a significant reduction in the size of the haystack may be welcomed.

There are, of course, serious ethical aspects to using this technology, including the violation of privacy, the possibility of the false identification of potential offenders and the stigma associated with such a false alarm. These ethical issues have solutions and they should be taken into account within a much broader perspective, which is *saving the lives of innocent people*. Whether we are trying to screen for a school shooter, a lone-wolf terrorist or a serial killer, we necessarily violate some civil rights. That being said, the idea of civil rights should not be dismissed as dogma, nor abused for wrongdoing. In this context, automatic personality analysis is an important tool (one among many others) that should and must be used for a just cause.

4.4   Separating the Barking from the Biting Dogs: Automatic Identification of Intentions

The analysis of texts written by school shooters and other perpetrators raises the question of how indicative texts are of individuals' *intentions*. Although this question leads us away from the main theme of this chapter, it is almost inevitable to ask it following the attempt to screen for potential offenders through computational personality analysis.

We can phrase this question in terms of barking versus biting dogs. The phrase "barking dogs seldom bite" suggests that people who make many threats seldom carry out those threats. How do we know whether violent and threatening language is an indication of real intentions? From a philosophical point of view, having an intention means that I conceive myself as an active agent who desires and strives to achieve a certain goal. I may wish death to my enemies but, unless I actively seek their death by having a plan, this death wish wouldn't be considered an "intention." From a legal point of view, judging a person as having a criminal intention means that a person must have had a plan of wrongdoing and have understood the *consequences* of his actions. In contrast with wishful thinking, in which a person generally wishes that something will miraculously happen in a rewarding way, intention, at least in the case of perpetrators, involves a deliberate plan to take active steps to achieve a harmful future goal while fully understanding its consequences.

Sometimes people respond to events with an uncontrolled burst of deadly rage. In other cases, clear intentions are expressed before an action is taken. For instance, the ISIL terrorists who committed the attack at the Bataclan theater in Paris in November 2015 had clear intentions. Therefore, it is highly important to develop tools to identify harmful intentions and to differentiate these intentions from wishful thinking per se and from empty "barking." Such an analysis should be done only as part of a risk-assessment procedure in which intentions are linked to capabilities. A 15-year-old Muslim extremist in Karachi who writes a death threat to the American president may be seriously expressing his anger and intentions, but these intentions are meaningless as they are probably disconnected from his real capabilities.

It is important to remember that, in contrast with how intention is perceived by some philosophical–mentalistic approaches, here we take intention to be an act of communication. When someone exposes his intentions, he does not merely reflect mental content but aims to communicate his desires and wishes. Therefore, and from the perspective of pragmatics, we should look for the contextual dimensions that help us to reconstruct the communicative intentions of a person. As argued by Bosco et al. (2004, p. 470), "an utterance [act of communication] extracted from its context of reference has no communicative meaning and cannot obtain any communicative effect." This statement calls into question the validity of the various approaches to computational intention analysis that are indifferent to the pragmatics of their analyzed content. Contextual dimensions of intention involve a deep understanding of background knowledge. For instance, let's have a look at the following utterance:

Wine?

Isolated from any contextual dimensions, this utterance is meaningless. However, adding one discursive move prior to this utterance may change the whole situation:

1. I'm so pleased to be playing host in my house.
2. Wine?

The first discursive move locates the second utterance in the context of hospitality and communicates our host's intention to serve us wine. We may use these contextual dimensions in order to design tools for the identification of intentions in textual data.

Bosco et al. (2004) suggest that we may study the context of intention communication by using dimensions such as access, space, time, discourse, move and status. The dimension of access concerns access to the physical object to which the communicative act refers (e.g. an object on which to carry out an action). For instance, in trying to identify the malevolent intentions of a teenager who might turn into a school shooter, we may ask whether there is a clear object at which intentions are targeted. A passionate Marxist student from South America may target his anger against global capitalism, believing it should be annihilated. As long as his anger is turned against an abstract and "faceless" object, we need not be much concerned. However, if his anger turns against a *specific object* with a face and a name, the changing context should increase our concerns. Another instance is the recent wave of terror in Israel during 2016, which is led, to a large extent, by angry Palestinian teenagers. A statistically significant number of these terrorist incidents have happened at the Damascus Gate, which leads into the old city of Jerusalem. This is not a coincidence, as Palestinian incitement on social media has continuously referred to this location in such a way as to turn it into a dimensional context of intention. This example leads us to the second dimension, which is space. Space concerns the spatial distance between agents and objects in the physical world to which the communicative act refers. When communicating their intention to act against Israelis and pointing to the specific nearby place and people to which their deeds are targeted, terrorists may strongly communicate the seriousness of their intentions. Automatic analysis of spatial language and geo-locations may be extremely helpful in this context. The dimensions of time, discourse, move and status can similarly be helpful in such analyses.

To illustrate the identification of intentions in a real text, let's examine the diaries written by the Columbine high school massacre perpetrators, starting with the depressive personality of Dylan Klebold. The Columbine high school massacre took place on April 20, 1999, and left 13 dead and 24 injured. This event is registered in the minds of American citizens as a trauma, and significant attempts were made to understand it. The two perpetrators, who were high school students, wrote diaries in which their inner worlds and personalities were exposed. When reading the text written by Klebold, we may identify a few sentences expressing intentions:

1. I want to be accepted.
2. Ever since X (who I wouldn't mind killing)….
3. I want to be with her.
4. I want to die sooo bad.

5. I want to find love.
6. Humanity is the something I long for.
7. I just want something I can never have.
8. I want pure bliss.
9. I'll go on my killing spree against anyone I want [his first reference to murder].
10. I want to be free.

Even if you are not a qualified psychologist, in reading these sentences, you may immediately understand that this is a depressive young man who suffers from social isolation and who is expressing aggressive intentions toward himself (as suicidal thoughts) and against others (as murderous intentions). However, in the diary from which these sentences were taken, we cannot find criminal intentions in terms of a clear object of desire, a carefully elaborated plan or any understanding of consequences. However, when analyzing the writings of Eric Harris—the second murderer—we are clearly exposed to a murderous mind. His intentions are clear, and here are just a few examples:

1. Hmmm just thinking if I want all humans dead.
2. I want to kill everyone.
3. I want to burn the world.
4. I want to tear a throat out with my own teeth.

In this case, we can see a clear reported intention to do harm. Moreover, a more detailed plan is elaborated than in Klebold's writing (e.g. "I will need a fucking fully loaded A-10"), and Harris displays an understanding of the consequences of his potential wrongdoing (e.g. "I know I can get shot by a cop... but hey guess the fuck WHAT! I chose to kill"). If we analyze verbs and nouns indicating intentionality (e.g. "want," "goal," "desire") in relation to the first-person pronoun, we immediately identify intentions to do harm associated with a disturbed belief system (e.g. "I feel like God"); themes of violence, including sexual violence; highly negative emotions (e.g. "I hate the fucking world"); and planned criminal behavior. Eric Harris, who was posthumously diagnosed as a psychopath, could have been investigated in advance if we could have had access to his recorded intentions and processed them automatically to send warning signals to the authorities. Having violent fantasies—even violent sexual fantasies—is not a justification for one's communications to be inspected. However, there is justification in taking preventive steps against a disturbed personality expressing vicious intentions with a close attention to the minor details of a plan (e.g. the chosen weapon) and an understanding of the consequences.

In sum, in this chapter, I have shown how effective the distributional semantics approach can be when applied to automatic personality analysis. This is a very simple approach that seems to work remarkably well in several contexts and that, surprisingly, even competes with some of the most sophisticated methods in the field. However, I have also pointed out how limited it is to focus computational personality on lower-level features or on semantic similarity only. Personality is about themes or areas of interest that occupy a person's mind, and the approach presented above is quite limited when used alone in understanding the themes and conflicts that constitute human personality. For instance, it is difficult to imagine how we may successfully identify defense mechanisms through distributional semantics only. The next chapter shows how we can extend the approach presented above by enriching our tool kit.

References

Bosco, F. M., Bucciarelli, M., & Bara, B. G. (2004). The fundamental context categories in understanding communicative intention. *Journal of Pragmatics, 36*(3), 467–488.

Davies, M. (2009). The 385+ million word Corpus of Contemporary American English (1990–2008+): Design, architecture, and linguistic insights. *International Journal of Corpus Linguistics, 14*(2), 159–190.

Knoll, J. L. (2010). The 'pseudocommando' mass murderer: Part I, the psychology of revenge and obliteration. *Journal of the American Academy of Psychiatry and the Law Online, 38*, 87–94.

King, S. (1974/2011). *Carrie*. London, UK: Hodder & Stoughton.

Langmeyer, A., Guglhör-Rudan, A., & Tarnai, C. (2012). What do music preferences reveal about personality? *Journal of Individual Differences, 33*(2), 119–130.

Millon, T., Millon, C. M., Meagher, S., Grossman, S., & Ramnath, R. (2012). Personality disorders in modern life. Hoboken, NJ: Wiley.

Mohammad, S. M., & Kiritchenko, S. (2013). Using nuances of emotion to identify personality. arXiv preprint arXiv:1309.6352.

Neuman, Y., & Cohen, Y. (2014). A vectorial semantics approach to personality assessment. *Scientific Reports, 4*, 4761.

Neuman, Y., Assaf, D., Cohen, Y., & Knoll, J. L. (2015). Profiling school shooters: Automatic text-based analysis. *Frontiers in Psychiatry, 6*, 86.

Saulsman, L. M., & Page, A. C. (2004). The five-factor model and personality disorder empirical literature: A meta-analytic review. *Clinical Psychology Review, 23*(8), 1055–1085.

Socher, R., Perelygin, A., Wu, J. Y., Chuang, J., Manning, C. D., Ng, A. Y., & Potts, C. et al. (2013, October). Recursive deep models for semantic compositionality over a sentiment treebank. In Proceedings of the conference on empirical methods in natural language processing (EMNLP) (Vol. 1631, pp. 1631–1642). Stroudsburg, PA: Association for Computational Linguistics.

Turney, P. D., & Pantel, P. (2010). From frequency to meaning: Vector space models of semantics. *Journal of Artificial Intelligence Research, 37*(1), 141–188.

Chapter 5
Themes of Personality:
Profiling a Political Leader

As consumers of mass media, we are recurrently exposed to more or less serious psychological profiles of political leaders. For instance, Vladimir Putin has been the subject of intense psychological interest (see, for instance, Robertson 2015) that has attributed to him various psychological dimensions including narcissism and autism. The interest in Putin and other powerful political leaders is clear. These people, specifically in non-democratic societies, hold enormous power, and understanding their personality may help us to better understand and hopefully anticipate their moves.

Intelligence agencies have their own profiling experts, and one of the most famous profilers is Jerrold Post, who has published a famous profile of the late Iraqi dictator Saddam Hussein as well as psychological profiles of other political leaders (Post 2010). In contrast with the psychological assessment of "ordinary" people, a political leader cannot be the object of a direct personality assessment. One cannot simply invite Vladimir Putin to participate in a psychological assessment day in which he fills out questionnaires, participates in interviews, and takes a part in group dynamics or projective tests (a pity as it could be interesting to hear Putin's interpretations of Rorschach inkblots...).

Given that direct assessment is impossible, there are two indirect methods of psychological assessment. The first approach is the more "impressionist," and involves assessment conducted "with the opinion of a single person, with standardized psychological assessment measures infrequently employed, and with official psychiatric diagnosis often ignored" (Coolidge and Segal 2009, p. 196). That is, the profiler is provided with all available data relevant to the personality assessment and attempts to provide a personality profile of the leader. For instance, if Western leaders meeting with Putin report that they feel he is behaving in a disrespectful and derogatory manner, a possible hypothesis is that he is a vain narcissist. In this context, the issue of *validity* presents a serious challenge to the personality assessment, which means that it is not clear how well our personality profile represents the *real* personality. For instance, cultural differences are a major source of potential bias. In Russian society, which is traditionally segregated, there is a huge gap between the elite and the common people. Individuals who hold

© Springer International Publishing Switzerland 2016
Y. Neuman, *Computational Personality Analysis*,
DOI 10.1007/978-3-319-42460-6_5

intellectual, political or other forms of high social status are expected to behave in a manner that expresses this status, and this manner might be conceived by outsiders as a derogatory form of vanity. Many Jewish Russian immigrants to Israel were previously part of the Soviet intellectual elite, and they have sometimes been accused by Israelis of being arrogant. In modern democratic societies with multi-party systems, having a high social, intellectual or political status cannot be practiced in the same way as in Russia (for example), as vain leaders run the risk of not being reelected. However, in less democratic and non-democratic societies, differences in social status are commonly manifested and practiced. Those who don't understand this unique cultural marker of Russian society cannot analyze Putin without confusing cultural differences with personality differences.

The second approach to indirect personality profiling employs informants' reports of others through the use of standard inventories. That is, in an attempt to be more structured and valid, it uses allegedly relievable and valid psychological tools of assessment. However, instead of delivering them to the leader to be completed, we ask those who know (or have known) the leader to fill them in (see Coolidge and Segal 2009). This approach suffers from severe difficulties, such as the extent to which others' reports regarding the subject may converge with one's self-report or even with an expert's perspective. Therefore, the validity of this approach is questionable.

How can we address these difficulties? Previously, I've put forward the idea of looking for themes of personality through textual analysis. In this context, though, this is clearly a problematic move. When the political leader himself generates a text, such as a diary written in his youth, we may gain some psychological insights from its analysis. But what happens if we seek to understand a political speech, which may have been written by others and/or be influenced by the politician's agenda? In this case, we don't gain any privileged access to the leader's mind per se but rather to a possible understanding of a *text in context*, where the context includes dimensions such as the way the text has been processed by the leader's close circle. But this is a perfectly legitimate approach. A leader would never deliver a speech that was not in line with his personality and policy, though it may have been written by a close assistant or edited by experts to polish its rhetorical impact. In this case, we should not be bothered by the question of who *really* wrote the speech, as we can assume that the question is not "Who is the author?" but "What is the *meaning* delivered through the text?" and "How much does the text represents the personality of the leader?" in its extended social and contextual sense. Understanding these points, we may move forward and see how the use of network motifs may help us in identifying themes of personality.

A text may be represented as a semantic graph of words or as arguments at a higher level. That is, we take a text that we would like to analyze and convert it into a graph comprising vertices (i.e. words or phrases) and edges (i.e. relations). For instance, in his 2015 speech to the UN assembly, President Putin said:

> We believe that any attempt to play games with terrorists, let alone to arm them, are not just short-sighted, but 'fire hazardous.'

We can see that President Putin clearly states his "beliefs" and that these beliefs can be represented as the relation between the collective "We" and the phrase dealing with the terrorists and the way they should or should not be handled.

These relations are significant. The methodology my colleagues and I developed in Neuman et al. (2015) translates a text into a graph by first automatically representing the sentences as a set of *binary relations* between words. The second phase involves the application of a set of rules that convert the dependency representation of the text into a graph with nodes/words and their *subject–object relations*. This semantic graph can be broken down into possible *sub-graphs* that comprise it. In this context, *network motifs* are recurrent and statistically significant sub-graphs. For instance, let's assume that we are interested in identifying three node sub-graphs. Let's tag these nodes as A, B and C. Now, there are various combinations in which these nodes may be linked to each other. These combinations are actually the potential motifs. For instance, A may send arrows to both B and C; B and C may send arrows to each other; and so on (Fig. 5.1).

When analyzing a network, we may want to learn about the building blocks from which it is comprised, and these building blocks can be considered to be the network motifs. Now the problem is that, when we partition a graph into all possible sub-graphs, we may get a huge number of sub-graphs that express all of the statistically possible combinations/configurations that can be produced by the partition. Here the idea of "motifs" comes into the picture. Motifs are sub-graphs that appear in a *significantly higher frequency* than what could have been expected by chance (Alon 2007). Through algorithms and specially developed software, these sub-graphs can be identified and processed further. In other words, there are algorithms that allow us to identify the statistically significant combinations (i.e. sub-graphs) of configurations comprising n nodes (e.g. three nodes) in a given graph.

Deciding whether a certain configuration appears in a graph in a statistically significant way is not a simple task, and several algorithms have been developed to address this challenge. Now, we may consider these motifs as *themes* that have some significance for understanding the personality that produced the text we are analyzing. I have proposed that, after identifying the motifs in a semantic graph, we should identify the most frequent words located at the motifs' nodes. In other words, by identifying the *words* comprising the nodes of significant motifs, we actually identify *themes* in the text—patterns of significant semantic relations. That is, we first identify the statistically significant configurations of objects and relations and then search for the most frequent words that populate the nodes of our motifs. It is very important to realize, though, that the most common method of motif analysis

Fig. 5.1 A network configuration

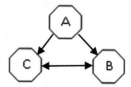

(as usually practiced in computational biology) is purely *structural* and focuses on the identification of statistically significant patterns of sub-graphs. This form of analysis is totally indifferent to the semantics of the nodes populating the motif. In contrast, traditional textual analysis is usually *semantic* and focuses on the "meaning" of signs or texts without taking into account statistical structural regularities of the words comprising the text. What is unique about the methodology I have developed (Neuman et al. 2015) is that it merges *form* (i.e. structural motifs) and *content* (i.e. semantics of nodes) in order to identify meaningful patterns of *relations* in a text. Now let us illustrate how this methodology can be used by analyzing a speech given by a political leader.

The speech is one given by the former Egyptian president Mohamed Morsi to the United Nations. Following the uprising in Egypt in 2011, the resignation of President Hosni Mubarak and the country's first democratic elections, Mohamed Morsi was elected to the presidency and served as the fifth president of Egypt from June 30, 2012, to July 3, 2013. His election represented an important change in the political map of the region and was of great concern to the United States, which has provided Egypt with billions of dollars in foreign aid over the years.

After many years in which the Muslim Brotherhood had been excluded from Egyptian politics, it decided to enter the lacuna created by Mubarak's resignation and to participate in the presidential election with Morsi as its candidate. Morsi's election raised some positive expectations in the United States and other Western governments, as he was considered to be "pragmatic," having received his PhD in engineering in the United States; it may have been expected that he would act favorably toward the country in which he had received his higher education. After he became president, Morsi's moves led to mass demonstrations among the Egyptian people, commencing in November 2012. On June 30, 2013 (the first anniversary of his presidency), Morsi was ousted by the Egyptian military, backed by a council consisting of defense minister el-Sisi, who became president in June 2014.

The speech my colleagues and I analyzed was given by Morsi on September 26, 2012, before the 67th session of the General Assembly of the United Nations. This speech attracted intense media coverage, not only due to the fact that this was the first speech given by an Egyptian president to the United Nations for more than a decade but also because this was the first speech given by a democratically elected president of Egypt. Moreover, just prior to the delivery of this speech, the region had been swept with anti-American propaganda, fueled by the release of an anti-Islamic movie considered by some to be an insult to the Prophet Muhammad. The movie triggered riots that included a violent attack on the American embassy in Cairo. The anti-American sentiments combined with the recent instability in Egypt got many world leaders worried about the stability of the entire region.

A "naive" and simple reading of the speech by my colleagues and I revealed five key themes/topics, such as the Palestinian–Israeli conflict and Egypt as an Arabic–African leader of the developing world. These themes are trivial in the sense that they could have been anticipated. However, what is interesting to find is what is hidden beneath the surface. In order to excavate hidden themes in Morsi's speech, my colleagues and I applied the methodology described above. This enabled the identification of several three-node motifs. After identifying the motifs and by automatically identifying the words populating most of the motifs' edges/nodes, three words emerged: "rights," "principles" and "vision." What is the significance of these three keywords?

Getting back into the text, we find that, when he refers to "vision," Morsi is mainly talking about a vision of Egypt, the new Egypt, a vision of Egypt's national security, the vision of the Egyptian people, and a vision of a steady democratic transfer of power. When talking about "rights," Morsi is mainly referring to the rights of the Palestinians. And, when he talks about "principles," we find keywords such as "law," "justice" and "righteousness." The big question is, "So what?" Here we move on to the interpretation phase, which is essential, as the automatic analysis may excavate interesting themes but it cannot interpret them for us.

Based on the analysis of the motifs, our own interpretation and hypothesis was that Morsi emerges from the speech as a man of principles and justice committed to the grand vision of a Muslim nation. From what you have learned so far, what can you say about personalities that are occupied by principles and justice? By doing the right thing? By a vision (i.e. a master plan) that must be executed? You may hypothesize, like we did, that Morsi's speech represents an obsessive–compulsive personality. As argued by Millon et al. (2012), obsessive personalities seek opportunities to prove that they are *selflessly committed to the "greater cause"* (i.e. vision in Morsi's speech). They are *rigid and dogmatic, occupied with details* and have an *overconscientious and scrupulous attitude toward morality and values*. This personality type seems to perfectly fit Morsi, the engineer who led the religious dogmatic organization of the Muslim Brotherhood.

Now, hypothesis production is not the bread and butter of computational personality analysis. However, a researcher, as well as a data scientist, works like a detective through a process of collecting cues, generating hypotheses and testing them empirically. Therefore, the second phase of our methodology involved the testing of our research hypothesis through the vectorial semantics approach, as presented in the previous chapter. To test the hypothesis that traits—or more accurately *themes*—of the obsessive personality appear in Morsi's speech, we surmised that, if indications of the obsessive–compulsive personality appear in the speech, the document should be semantically similar to the vector of words characterizing the *Diagnostic and Statistical Manual of Mental Disorders'* definition of the obsessive–compulsive dimension. The logic of our approach was that expert psychologists and psychiatrists can characterize personality types by using minimal sets of words that grasp the essence of the personality. For instance, for the paranoid personality, the adjective "suspicious" is a prototypical keyword. Using a set of adjectives that describe a certain personality, we may automatically analyze the dimensions of a given text by simply representing it as a vector and measuring its similarity to the vector formed from the words that characterize the personality. In our case, we identified a set of words describing obsessive personality and measured the similarity of the vector composed of these words to the vector produced from Morsi's speech.

However, a similarity score in itself is meaningless. It is just a number. To test the hypothesis that Morsi's speech expresses certain personality characteristics, we produced the following competing hypothesis: that, instead of displaying an obsessive–compulsive personality, Morsi's speech is semantically similar to the vector of words characterizing the narcissistic personality—a personality type many leaders are suspected of exhibiting. In support of our original hypothesis, however, we found that the vector of the speech was much closer to the obsessive–compulsive personality vector than to the narcissistic personality vector. In order to reach a more specific "diagnosis," we analyzed the speech's similarity to the vectors of Millon and colleagues' five sub-types of compulsive personality (Millon et al. 2012). The vectors of the five sub-types are as follows:

A. *Conscientious*: rule-bound, duty-bound, earnest, hard worker, meticulous, indecisive, inflexible
B. *Bureaucratic*: officious, high-handed, unimaginative, intrusive, nosy, petty-minded, meddlesome, trifling, closed-minded
C. *Puritanical*: austere, self-righteous, bigoted, dogmatic, zealous, uncompromising, indignant, judgmental
D. *Parsimonious*: miserly, tight-fisted, ungiving, hoarding, unsharing
E. *Bedeviled*: ambivalent, tormented, muddled, indecisive, befuddled, confused, frustrated, obsessed.

We found that the speech was most similar to the bureaucratic and puritanical sub-types (0.09 and 0.10 respectively). No or very low similarity was found with the other sub-types. Based on these findings, we diagnosed Morsi's personality, at least as reflected in this specific speech, as consistent with the dimension of the obsessive personality, reflecting a *bureaucratic–puritanical leader*.

As suggested by Millon et al. (2012), one of the main pitfalls of the obsessive personality is a failure to see the big picture. This suggestion seems to contradict Morsi's themes of vision. However, in retrospect, this pitfall clearly explains Morsi's failure to gain the support of his people and to solve the deep problems of Egyptian society. Specifically, Morsi's ignorance of the socio-political situation in Egypt; his focus on the bureaucratic–puritanical issues of the new regime; and his desire to execute a master plan that did not address the deep concerns of the Egyptian people together led to his collapse. The speech, whether written by Morsi himself or—more plausibly—by Morsi and his close circle, represents strong features of the obsessive–compulsive personality, characterizing a puritan who holds himself to the rigid moral norms of his religious community in the same way as he holds himself to the bureaucracy of the Muslim Brotherhood. It would have been very difficult for such a leader to employ transformative actions consistent with the urgent needs of his society. In retrospect, this analysis explains the failure of Morsi and his tragic political end as a prisoner, as of 2016 awaiting a death sentence.

While our analysis extends the distributional semantics approach presented in the previous chapter by including analysis of network motifs, it is still a limited methodology. Human communication takes place at the *discursive level of analysis*, at which we can grasp the pragmatics of human communication. In linguistics, "pragmatics" is defined as the use of language in human communication as determined by the conditions of society (Mey 1993). Understanding the "pragmatics of personality" is probably the next challenge facing computational personality analysis, and in this context developing novel methodologies is a must. The next chapter presents one such methodology that I have recently developed (Neuman and Cohen, in press).

References

Alon, U. (2007). Network motifs: Theory and experimental approaches. *Nature, 8,* 450eU.

Coolidge, F. L., & Segal, D. L. (2009). Is Kim Jong-il like Saddam Hussein and Adolf Hitler? A personality disorder evaluation. *Behavioral Sciences of Terrorism and Political Aggression, 1,* 195–202.

Mey, J. L. (1993). *Pragmatics: An introduction.* Oxford, UK: Blackwell.

Millon, T., Millon, C. M., Meagher, S., Grossman, S. & Ramnath, R. (2012). *Personality disorders in modern life.* Hoboken, NJ: John Wiley & Sons.

Neuman, Y. & Cohen, Y. (in press). A novel methodology for extracting psychological dimensions from textual data. *Computer Journal: C.*

Neuman, Y., Cohen, Y., & Shahar, G. (2015). A novel computer-assisted personality profiling methodology. *American Intelligence Journal, 32*(1), 136–146.

Post, J. M. (Ed.). (2010). *The psychological assessment of political leaders: With profiles of Saddam Hussein and Bill Clinton.* Ann Arbor, MI: University of Michigan Press.

Robertson, I. H. (2015, February 24). Inside the mind of Vladimir Putin. *The Telegraph.* http://www.telegraph.co.uk/news/worldnews/europe/ukraine/11431850/Inside-the-mind-of-Vladimir-Putin.html. Accessed April 22, 2016.

Chapter 6
Going Beyond Words: How the Sisters of Mercy May Identify Psychopaths

In the preceding chapters, I've explained that computational personality has usually been addressed as a specific instance of an ML-based classification. Studies have used corpora of texts tagged with the personalities of their authors, and the success of classifiers has been examined. One problem with these previous studies is that they have focused on words rather than on higher-level features that may be indicative of the more complex discursive aspects that characterize human personality. This problem was discussed in the previous chapter. However there is another problem, which is the availability of tagged corpora. It is quite difficult, even in the age of Big Data and Amazon's Mechanical Turk, to gain a high-quality corpus that includes, on the one hand, texts that have been "naturally" produced and, on the other hand, the measured psychological dimensions of those texts' authors. This problem is painfully evident when we try to find a high-quality corpus of psychological dimensions that are relatively rare. For example, if we would like to build a classifier that screens people who are psychopaths, then, given the very low prevalence of psychopaths in the population, less than 1 %, gaining a big enough and high-quality training corpus for our classifier may be very difficult if not impossible.

I keep emphasizing the importance of having a *high-quality corpus* because some of the corpora used in the field of computational personality are far from perfect, to say the least. The corpus of essays that is frequently used in computational personality workshops (e.g. Celli et al. 2013) as well as in my studies is very problematic. I have done what ML experts usually don't—I have read some of the essays comprising the corpus. Let's put aside the fact that these essays were written by a non-representative sample of students and therefore cannot be considered to be a representative sample of any population, and let's put aside the fact that these are not naturally produced essays but the product of an experimental artificial manipulation. Even then, the essays themselves seem to have low-quality content that is difficult to decipher for a valid personality analysis. To make a long story short, getting your hands on a high-quality corpus for personality analysis might be a pain in the ass. In this context, I've proposed a solution that hasn't been presented or

© Springer International Publishing Switzerland 2016
Y. Neuman, *Computational Personality Analysis*,
DOI 10.1007/978-3-319-42460-6_6

tested before. This solution is described in a paper (Neuman and Cohen, in press) and detailed below. This solution addresses both the availability problem and the need to identify higher-order features in texts.

I have previously mentioned the fact that traditional psychology has usually relied on in-depth interviews and questionnaires in order to diagnose human personality. I have also mentioned the difficulties of this approach when we seek to analyze a massive amount of textual data. However, a lot of effort has been invested in developing valid psychological questionnaires. My new idea is to design a procedure in which a psychological questionnaire or inventory is administered to a written text rather than to an individual. This idea may have several clear advantages. First, when delivering a psychological questionnaire to an individual, the individual is asked to rate the extent to which certain personality items describe him. This is problematic in several senses, one being that the person is asked to be highly reflective and honest, two challenging demands for *Homo sapiens*. By "delivering" the questionnaire to naturally written texts—that is, texts written without any experimental trigger—we don't have to assume that the subject who wrote the texts is reflective and we don't have to be worried that he is lying to us. In addition, we may analyze a huge number of texts written by a single person and therefore gain information that extends beyond a specific context. The second gain in using my proposed methodology is that it enables us to automatically analyze a massive number of texts using a procedure that goes beyond low-level features. Let me be more specific about the methodology by using an example, one among quite a few I'm going to use in this chapter.

Let's assume that an imaginary Christian organization titled the Sisters of Mercy is recruiting new volunteers among college students to aid a developing country. The great news is that volunteering for the one-year mission will be compensated by a two-year scholarship to study in the United States. As expected, the Sisters of Mercy receive thousands of applications from which to choose, and they employ an expert psychologist to assist them. Their first decision is to reduce the number of potential candidates by rejecting people who are characterized by unwanted psychological traits such as psychopathy, as such people lack empathy and remorse. They ask the expert to screen for and exclude all of the candidates who have high scores on a valid measure of psychopathy. The expert psychologist suggests the use of the Levenson Self-Report Psychopathy Scale (Levenson et al. 1995), which includes items such as:

I often admire a really clever scam.

Sister Mary, who is in charge of the recruitment process, immediately identifies one major flaw with this suggestion, which is that an individual would have to be quite stupid to report strong agreement with the above item when asking to join a humanitarian mission. Sister Mary considers a second option, put forward by a second expert, which is the use of ML for automatic classification. The ML expert advises her to build a training corpus of tagged texts written by psychopaths and non-psychopaths and to train a classifier using this corpus. This classifier will be applied to a corpus of essays written by the candidates in order to identify those that seem to express dimensions of psychopathic personality. However, the problem with this approach is that such a training corpus is unavailable and, given the low prevalence of psychopaths in the population, building such a corpus from scratch might be very difficult. However, sister Mary recalls that she recently heard of a paper by Professor Neuman that offers another solution. Here is the proposed solution to the problem of screening for psychopaths.

Today, many people "live" in social media. It would be perfectly legitimate to ask candidates for the mission to provide the Sisters of Mercy with access to their Facebook page, just like they would grant access to their friends. After receiving access and authorization to process the data, the textual data of each candidate are downloaded and the sentences comprising each text are preprocessed. Next, we identify the sentences in which the first-person pronoun "I" is associated with a verb or a noun phrase. For example, let's assume that Adolph is a young candidate for the Sisters' mission. Searching his Facebook pages, we find the following sentence:

> I enjoy seeing people being cheated.

This sentence may be considered to indicate a lack of empathy and therefore it may be a sign of psychopathy. The question is, how do we decide whether this sentence may be indicative of psychopathy? We may measure the *semantic* similarity between the item:

> I often admire a really clever scam.

and the sentence:

> I enjoy seeing people being cheated.

If these sentences are similar in their *meaning*, we can raise Adolph's score on our psychopathy index. By using this approach, we may produce, from each text, measures indicating the degree to which the target psychological dimension is evident.

To better clarify this point, let's use another example that appears in our paper (Neuman and Cohen, in press). This time we would like to measure the personality trait of extraversion, and we use the following item, drawn from a personality questionnaire:

> I like people; I am friendly and open talking to strangers.

This item may be represented through the following core propositions:

1. I like people.
2. I am friendly.
3. I am open talking to strangers.

The proposition "I like people" can be represented as LIKE[I, PEOPLE] and in terms of dependency parsing it can be represented as follows:

nsubj(like-2, I-1)

root(ROOT-0, like-2)

dobj(like-2, people-3)

This dependency representation of the item's first proposition is embedded in a procedure that aims to measure the similarity between a target sentence and our item's propositions. For the above example, and for a given target sentence that has the formal structure of nsubj(VB, I) dobj(VB, NN), this procedure is as follows:

IF nsubj(VB, I) dobj(VB, NN)

THEN

SIM(VB, LIKE) SIM (NN, PEOPLE)

IF Advmod (VB, RB = "seldom" or "rarely" or "barely" or "hardly" or "infrequently" or

"never" or "scarcely" or "almost never" or "not often")

OR

IF neg (VB)

THEN SIM = SIM*(-1)

Let me explain this procedure in simple words. Assume that you are analyzing a text where a person writes: "I love human beings." This sentence can be represented as:

nsubj(love-2, I-1)

root(ROOT-0, love-2)

dobj(love-2, human-beings-3)

Therefore, we may measure its similarity to the proposition "I like people" by measuring the semantic similarity between "like" and "love" and between "people" and "human beings," and by taking into account the possibility of negation (i.e. "I don't love human beings") or qualification (e.g. "I rarely love human beings"). In sum, the proposed methodology for measuring a psychological dimension in a text is as follows:

1. First, select relevant items from a questionnaire that validly measures your personality dimension. You don't have to use all of the items as some questionnaires are quite long and their items may be redundant.
2. Identify the propositions comprising each of your items.
3. Translate your propositions into lexico-semantic patterns using a syntactic parser.
4. Process your target text and identify in it relevant sentences for analysis.
5. Design micro-procedures to measure the *semantic similarity* of the relevant sentences in a target text to the proposition extracted from the item.

Let's illustrate these steps, this time while trying to measure neuroticism in a target text. The item we use to measure neuroticism is:

I often feel nervous, fearful and anxious.

The core propositions extracted from this item are:

FEEL[I, NERVOUS]

FEEL[I, FEARFUL]

FEEL[I, ANXIOUS]

And here is the measurement procedure:

IF nsubj (X, I) AND IF X = JJ

THEN SIM(X, nervous/fearful/anxious)

IF advmod (X, RB = "seldom" or "rarely" or "barely" or "hardly" or

"infrequently" or "never" or "scarcely" or "almost never" or "not

often")

OR

IF neg(X)

THEN SIM = SIM*(-1)

Now for the explanation.

- *Lines 1–2*: If, in a sentence identified in our target text, the word associated with "I" through the nsubj is an adjective, we should measure the similarity between this adjective and the vector of the three words "nervous," "fearful" and "anxious."

- *Lines 3–6*: However, if the word associated with X, which is an adjective, is an adverb belonging to one of the words in the list (e.g. "seldom," "anxious"), then turn the similarity score into a negative score indicating non-neuroticism. The same is true for the negation of X (e.g. "I'm not anxious").

Here is an another example, this time of measuring narcissism through the Narcissistic Personality Inventory (Raskin and Terry 1988). Let's assume that you are a profiler working for an intelligence agency. You are given a mission to analyze the personality of a politician who is a promising candidate to lead a friendly country. You are asked to build a profile of your target and—most surprisingly—get access to personal letters and diaries this politician wrote between his childhood and his college days. As politicians are usually accused of being narcissists, you ask yourself whether there is a structural way to measure narcissism in the politician's texts. You use the Narcissistic Personality Inventory, in which you find the following item:

I think I am a special person.

By using automatic dependency parsing, you get the following representation:

nsubj(think-2, I-1)

root(ROOT-0, think-2)

nsubj(person-7, I-3)

cop(person-7, am-4)

det(person-7, a-5)

amod(person-7, special-6)

ccomp(think-2, person-7)

Next, when you search the politician's "confessional" diary, you find the following sentence:

I am better than anyone else.

You may find the semantic similarity between these two sentences to be quite high. And, given a representative corpus for comparison (e.g. a corpus of blogs drawn from the same population), you may reach some interesting conclusions about your target in a way that is much more empirically grounded and persuasive than a purely impressionist psychological diagnosis.

The similarity scores produced by the micro-procedures outlined in Neuman and Cohen (in press) are used in order to produce a global score indicating the extent to which a psychological dimension is evident in a text. We tested the methodology in several contexts and found it to be highly effective. For example, we used it to measure the level of depression in texts and used this level of depression to identify the school shooters in the test corpus described in Chap. 4. It was found that using the depression score was much more effective in screening the shooters' texts than the more complex features and methodologies we used in our study of the shooters.

Along the same line of reasoning presented above, the proposed pragmatic shift in the computational study of personality may be illustrated through an interesting study that attempted to uncover the "persona" of film characters (Bamman et al. 2013). In this paper, the researchers used plot summaries of movies that included concise descriptions of the characters. The processed the texts, extracting three main sources of features that relate to personality: agent verbs, which concern actions taken by the character (e.g. "He *ran* away"); patient verbs, which concern actions done to a character (e.g. "He was *stabbed* to death"); and attributes of a character (e.g. "a nasty fellow"). Then, they defined the persona of a character as the distributions of these three sources of features. In this sense, the persona of a character is actually the typical actions and attributes associated with it. Using this idea, the authors were able to infer the latent persona of film characters with success. It seems that we can use the same idea to recognize the personality of real human beings by looking for patterns in the actions they take, in actions taken upon them and their (self-) descriptions. It would also be possible to combine this methodology with the methodology proposed above to find out how similar one's personality is to the personae of different film characters.

So what are the pros and cons of the methodology I've proposed? The pros are clear, as they were the initial motivation to develop this methodology: measuring personality dimensions in naturally produced texts without a tagged corpus and by taking into account the pragmatics of human personality and its expression in more complex linguistic levels of analysis. The major cons are, first, the need for human expertise to produce patterns and micro-procedures, and, second, the fact that the similarity procedure is not quite simple to apply. The first difficulty might be over-exaggerated as it took me only a few hours of work to formulate the micro-procedures. However, the second difficulty is more concerning. To explain this problem, let us return to the example of measuring narcissism in a text. Given the personality item:

I think I am a special person.

we find in a text the following sentence:

God loves me as if I were his own child.

Now, in a certain context this sentence doesn't have to be indicative of a narcissistic personality. If I am a Baptist preacher in the American South who is deeply loved by my community, using this sentence in a religious ceremony may make some sense and doesn't have to be indicative of my personality. However, if I am a politician who produces this sentence during a campaign, this statement may be indicative of my personality.

Regardless of the context issue, it seems reasonable to deduce that if (1) God loves me as if I were his own child then (2) I think I am a special person. However, proving this textual–psychological link is impossible using our current methodology, with its limited number of patterns. A future development in computational personality analysis will allow us to use a predefined set of personality items and an engine that validly measures the extent to which our personality items are psychologically *entailed* by a given text. Before developing the methodology described in Neuman and Cohen (in press), I experienced with several tools for entailment, and the results were quite disappointing. For example, I used the Probabilistic Lexical Inference System (http://irsrv2.cs.biu.ac.il/nlp-net/PLIS.html) and checked that if this sentence:

I am self-critical.

is derived from the hypothesis:

I blame myself for everything that goes wrong.

For every armchair psychologist, it is clear that, if you blame yourself for everything that goes wrong, you are self-critical. However, the system described above scored the probability of there being an inference between the two statements as only 2 %.

Let's further elaborate the idea of a personality entailment engine by using the beginning of a transcript in which a client discusses the strain on her marriage caused by mistrust, suicidal tendencies, hospitalizations and a dependence on therapy (A, n.d.).

When examining propositions associated in this text with the first-person pronoun (i.e. I), we find that the patient presents three main informative pieces of information:

1. I want to take care of Jim [her husband].
2. Jim thinks I'm flaky, untrustworthy and incapable.
3. Jim is right.

Now, the conclusion is that "Ma"—the patient—believes that she is flaky, untrustworthy and incapable—but how can we automatically derive this important conclusion from her text? This move is computationally far from trivial and we may seek a simple approach to producing this valuable information. The approach may be to ask in a top-down manner whether certain indications of known personality dimensions are evident in a text. Let's take self-criticism, for instance. Self-criticism, self-hate, self-disgust, self-devaluation, self-condemnation and so on are experiences of oneself as a failure. While self-criticism can have a positive aspect, beyond a certain corrective level it is associated with feelings of inferiority and inadequacy to the level of self-hate and thoughts of self-punishment. It is clear, therefore, why self-criticism is associated with depression and suicidal thoughts. How can we measure self-criticism in a text? We may apply the following steps:

1. First, we may identify canonical propositions and sentences expressing the dimension that we would like to measure. For instance, "I'm a failure," "I'm very irritable when I have failed," "I'm disappointed with myself." In addition, we may identify indirect indications of self-criticism such as calling oneself names ("What a stupid asshole I am!").
2. Next, we may paraphrase each proposition either using platforms such as Amazon's Mechanical Turk or using an automatic paraphrasing engine that is based on (for instance) natural logic inference, such as the one proposed by Angeli and Manning (2014).
3. We search a relevant corpus for the canonical propositions and their paraphrases.
4. In each text, we identify the context in which our target proposition appears.
5. We build a semantic–syntactic directed and weighted graph of all propositions gathered.
6. We provide an abstract version of the graph by merging edges/vertices that provide redundant information and experiment with various levels of abstraction.
7. We analyze the target text for indications of the personality dimension, producing a graph and measuring how similar it is to the graph previously constructed.

In sum, in this chapter, I've proposed that we should move forward from the analysis of low-level features to the analysis of propositions. While low-level features, such as n-grams, have been remarkably effective in various tasks of natural language processing, such as in the context of authorship attribution (e.g. Koppel et al. 2011), it seems that they have their own limits for tasks in which we would like to fully represent and understand the complexity of human personality. In this context, the development of what I've described as an engine for psychological entailment may be an important development in the field and an economical way of processing texts to measure a wide variety of psychological dimensions from self-criticism to aggression management.

References

A. (n.d.). Client "Ma," session March 24, 2014: Client discusses the strain on her marriage due to mistrust, suicidal tendencies, hospitalizations, and a dependance on therapy. In *Counseling and psychotherapy transcripts: Volume II*. Alexandria, VA: Alexander Street.

Angeli, G. & Manning, C. D. (2014, October). NaturalLI: Natural logic inference for common sense reasoning. In *Proceedings of the 2014 Conference on Empirical Methods in Natural Language Processing, Doha, Qatar, 25 to 29 October 2014* (pp. 534–545). Stroudsburg, PA: Association for Computational Linguistics.

Bamman, D., O'Connor, B. & Smith, N. A. (2013). Learning latent personas of film characters. In *Proceedings of the 51st Annual Meeting of the Association for Computational Linguistics* (Vol. 1: *Long Papers*, pp. 352–361). Stroudsburg, PA: Association for Computational Linguistics.

Celli, F., Pianesi, F., Stillwell, D. & Kosinski, M. (2013, June). Workshop on computational personality recognition (shared task). In *Proceedings of the Workshop on Computational Personality Recognition*. Boston, MA: AAAI Press.

Koppel, M., Schler, J. & Argamon, S. (2011). Authorship attribution in the wild. *Language Resources and Evaluation, 45*(1), 83–94.

Levenson, M. R., Kiehl, K. A. & Fitzpatrick, C. M. (1995). Assessing psychopathic attributes in a noninstitutionalized population. *Journal of Personality and Social Psychology, 68*(1), 151–158.

Raskin, R. & Terry, H. (1988). A principal-components analysis of the narcissistic personality inventory and further evidence of its construct validity. *Journal of Personality and Social Psychology, 54*, 890–902.

Chapter 7
Hidden Textual Themes: Into Shakespeare's Mind

Computational tools may be used not only to help us in the task of classification but also as *exploratory* tools that may support the serendipity that necessarily accompanies an authentic scientific journey. I don't use the "exploratory" in the sense of "exploratory data analysis" but in the sense of trying to excavate hidden meaning in a text through structured and automatic "micro-reading" of a text. Let me explain this point by starting with good old Dr. Freud.

Freud's objective for scientific psychology was naturally inspired by the scientific advancements of his time and by his dream of bridging the gap between "mental functioning" and the "nervous system." That is, Freud's original vision was in line with what we currently observe in neuroscience. However, for Freud it was speech, or more specifically the talking (or writing) subject, that was the ultimate source of information and not the nervous system of this individual. The vicissitudes of life are such that Freud's vision can now be realized by using computational tools to find patterns in various systems of meaning, from written to spoken texts. In this chapter, I would like to introduce the idea of "computational psychoanalysis," which draws on psychoanalytic theories and computational tools for understanding the human mind. This is not an established field of inquiry, although it is evident in some studies, such as those by my colleague Fionn Murtagh (e.g. Murtagh 2014, 2016). More specifically, I draw on several ideas introduced by the psychoanalyst Matte-Blanco (2012) in order to show how computational psychoanalysis may illuminate deep layers of textual data. In contrast with the fairly conservative attitude adopted throughout most of this book, this chapter plays on the wild side …

Matte-Blanco's main theoretical thesis is that *symmetry* is the hallmark of the *unconscious* (Matte-Blanco 2012). Simply stated, symmetry is the process through which identity is established between elements of the mind through certain psychological transformations. For instance, if I conceive both my boss and my father as authority figures, my boss is symmetrical with my father to a level in which they may be equated and confused with each other. This idea may be quite acceptable among psychologists following the psychodynamic approach. After all, having a

© Springer International Publishing Switzerland 2016
Y. Neuman, *Computational Personality Analysis*,
DOI 10.1007/978-3-319-42460-6_7

problem with authority figures stretches across the board from parents to teachers and supervisors.

However, is it really the case that my boss *is* my father, in the strong sense of identity? Matte-Blanco's concept of symmetry draws heavily on a very limited notion of symmetry in set theory, but symmetry has much more to offer. Let me suggest that my mind, rather than equating my boss with my father through their identities, is *relating* my boss with my father through analogy or metaphor. That is, instead of reasoning that my supervisor *is* my father, a probable but quite disturbed psychological conclusion, it seems that my mind is somehow reasoning that my boss *is like* my father. In this case, my father cannot simply be exchanged with my boss, and the issue is not one of symmetry but one of *analogical or metaphorical reasoning*. What does this mean?

A metaphor/analogy is way of comparing two objects by "mapping" deep relations from one domain to another. Saying, for instance, that my boss *is like* my father may metaphorically mean that my boss patronizes me the same as my father patronized me. Figure 7.1 represents this graphically. In fact, deducing the abstract idea that both my boss and my father are "authority figures" is possible only by acknowledging that both figures "factor" through each other with regard to the "patronization" relation in which I am the object. This relation is illustrated in Fig. 7.1 through the horizontal bi-directional arrow connecting "father" and "boss." It means that we can equate my boss and my father, as both of them share the relation of patronization with "me."

Therefore, identifying analogical (a is to b like c is to d) or metaphorical relations (a is like b) may illuminate the unconscious, hidden layers of a text, a point that I would like to illustrate through the analysis of one of Shakespeare's sonnets. However, let me start with a formal computational analysis of analogy.

Let's take the analogy "a surgeon is like a butcher." We may explain the analogy by saying that the surgeon cuts patients' flesh in the same way as the butcher cuts meat (animals' flesh). Symbolizing the surgeon as "a," the butcher as "c," flesh as "b" and meat as "d," the analogy is symbolically represented as follows:

a:b::c:d

surgeon:flesh::butcher:meat

My colleague Turney (2013) has made an interesting proposal regarding the way we analyze analogies through the simple semantic similarity between words.

Fig. 7.1 The analogy father–boss

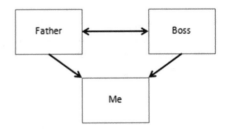

Fig. 7.2 The structure of
analogy

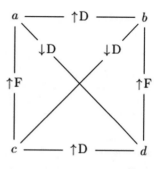

$$\text{sim}_r(a:b, c:d)$$
relational similarity

Turney differentiates between *domain similarity* and *function similarity*. Domain similarity concerns the extent to which two objects belong to the same semantic field. For instance, "butcher" and "meat" have high domain similarity, as both of them belong to the semantic field dealing with the consumption of meat as a food. In contrast, function similarity concerns similarity of *relations*. For instance, take the analogy "traffic:street::water:riverbed." Traffic–street and water–riverbed are two word pairs that have internal domain similarity, but traffic–water has a high function similarity as both of them "flow" through a certain medium, whether "pipe" or "street." To find a good analogy, Turney suggests using the idea illustrated in Fig. 7.2.

Simply stated, the idea is that, in analogy, we maximize domain similarity (↑D) and function similarity (↑F) between certain objects of an analogy while minimizing domain similarity (↓D) between other objects. Turney's idea can be illustrated with regard to a simple analogy test, such as the following:

"Fish" is to "fin" the same as "bird" is to…

1. beak
2. feather
3. wing
4. claw

Let us represent this task as fish:fin::bird:c, where "c" signifies the correct answer. In "teaching" the computer to choose the correct answer, we will guide it to work according to the above-mentioned model and to choose the answer that maximizes the domain similarity between fish and fin and between bird and "c"; minimizes the domain similarity between fish and bird and between bird and "c"; and maximizes the function similarity between fish and bird and between fin and "c". This is an idea that is far from trivial and that will be made much more clear through the analysis of a concrete example. So, let us move on to Shakespeare's famous sonnets.

The sonnets have an interesting rhyme scheme, which is abab cdcd efef gg. Let us read together the first four stanzas of Sonnet 1, which I have analyzed in a paper published in *Semiotica* (Neuman 2013).

> From fairest creatures we desire **increase**,
> That thereby beauty's rose might never *die*,
> But as the riper should by time **decease**,
> His tender heir might bear his *memory*:

As you can see, the sonnet is organized according to a well-defined structure that involves rhymes (articulatory-acoustic relations between syllables—e.g. increase–decease). These rhymes are actually *repetitions* that form analogies, because, if "decease" acoustically echoes "increase," then an association, and as I have elsewhere argued an analogy, is formed in our mind. If the analogy is a form of symmetry, which according to Matte-Blanco exposes unconscious materials, then automatically identifying analogies/metaphors in a text may be a powerful tool in exposing hidden unconscious content!

Let us move on and try to understand the psychological meaning of one analogy in Shakespeare's sonnets. Here, I would like to focus on one analogy emerging from his rhyme structure, which is "womb is like a tomb." The metaphor is illustrated in Sonnet 3:

> [1] Look in thy glass, and tell the face thou viewest
> [2] Now is the time that face should form another;
> [3] Whose fresh repair if now thou not renewest,
> [4] Thou dost beguile the world, unless some mother.
> [5] For where is she so fair whose unear'd **womb**
> [6] Disdains the tillage of thy husbandry?
> [7] Or who is he so fond will be the **tomb**.

What does it mean that the "womb" is like a "tomb"? Analogically speaking, it means that:

> Womb:b::Tomb:d

but how can we find the missing arguments of the analogy—that is, "b" and "d"? Finding possible candidates for "b" and "d" isn't easy. However, one possibility is to search for words collocated with "womb" and "tomb" in a large linguistic corpus. For this task, I used the Corpus of Global Web-based English (GloWbE; http://corpus2.byu.edu/glowbe), which contains 1.9 billion words. I searched for nouns and verbs collocated four words to the right or left of "tomb" and specified that their collocation with the target words should be significantly "informative." Figure 7.3 shows several results found for "tomb," and Fig. 7.4 shows results gained for "womb".

Now, these results and many more we may retrieve from the corpus allow us to construct various combinations of words that generate propositions, even nonsensical propositions such as that "mother is formed in the womb." However, seeking the analogy between "womb" and "tomb," I first measured the semantic similarity between the verbs/relations associated with "womb" and those associated with

Fig. 7.3 Collocations of "tomb"

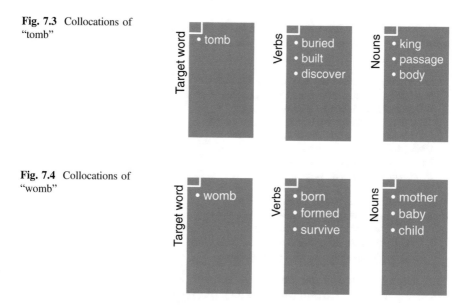

Fig. 7.4 Collocations of "womb"

"tomb." Using a common automatic device for measuring semantic similarity, I found that the strongest semantic similarity exists between the verbs:

buried and **formed**

Next, I used Turney's model described above and selected the best "b" argument for "womb" from among the three options: "mother," "baby" and "child." We find that the argument that has the greatest similarity with "womb" and the least similarity with "tomb" is "baby." Therefore, we have the first part of the analogy:

baby formed [in the] womb

Applying the same procedure for "tomb," we find the most similar argument is "king," so:

king [is] buried [in] tomb

The final analogy is:

the baby is formed in the womb like the king is buried in the tomb

The analogy we have just formulated is just one of several possible hypotheses that we can automatically produce through the above-mentioned procedure. In other words, our analysis first generates several possible analogies, but the most appropriate analogy is selected according to the model proposed by Turney. This automatic procedure excavates hidden analogies in the text, with the hope that these hidden analogies can teach us something interesting about the unconscious of Shakespeare. However, there is no substitute for the human expert's interpretation.

Now we should interpret the meaning of the analogy and see how helpful it is in illuminating certain hidden aspects of the text and the poet's mind and personality.

Here is Sonnet 3 in its entirety:

> Look in thy glass, and tell the face thou viewest
> Now is the time that face should form another;
> Whose fresh repair if now thou not renewest,
> Thou dost beguile the world, unbless some mother,
> For where is she so fair whose unear'd womb
> Disdains the tillage of thy husbandry?
> Or who is he so fond will be the tomb
> Of his self-love, to stop posterity?
> Thou art thy mother's glass, and she in thee
> Calls back the lovely April of her prime:
> So thou through windows of thine age shall see
> Despite of wrinkles this thy golden time.
> > But if thou live, remember'd not to be,
> > Die single, and thine image dies with thee.

A literary reading of the sonnet is as follows. The poet is presenting an interesting argument—the reason why his young friend should start a family. The young man need only look at his own mother to see how important his youth and beauty are to her, as a constant reminder of "the lovely April of her prime." The poet urges the young man to let another woman experience this joy, by having his children, and he too will benefit from seeing in his child his former "golden time."

This interpretation is strikingly *opposed* to the analogy we have just found through the analysis of the womb–tomb analogy; if bringing a child into the world is such a happy experience, why is the womb compared to a tomb? In fact, following the logic of the analogy we've just identified, we may say that the baby/king who is formed in the womb is actually *buried* in it as if it were a grave! Given this chain of reasoning, the mother, who is ideally described as experiencing the pleasure of giving life, is actually exposed through our computational analysis as a destructive force that annihilates the baby/king. As in the field of psychological symbolism, the king is the ultimate expression of narcissism, being a source of admiration, and we may read the analogy we have just identified as follows:

> The king/baby is experiencing an infantile narcissistic sense of importance. However, as his mother fails to satisfy his narcissistic demands, *he splits her* (see Chap. 2) into good and bad mothers. The bad mother is not a source of life but a graveyard where the baby/king is doomed to be buried in her womb/tomb.

This reading of Shakespeare's sonnet suggests that, while on surface (on the conscious level) Shakespeare is portraying the mother and motherhood in ideal positive terms, this description is actually highly ambivalent and unconsciously involves a deep and dark conception of the mother as *an annihilating force*. For psychoanalysts this conclusion may not come as a surprise. When one digs deeply into the unconscious, one usually finds skeletons rather than treasures. However, it was the implementation of Matte-Blanco's ideas through novel computational tools that led us to the above conclusion, which involves a radical and critical reading of

Shakespeare's sonnet in a way that challenges a more straightforward interpretation of the text.

What can we learn about Shakespeare's personality through the above-mentioned analysis? The answer is that, through one analyzed analogy, nothing can be learned except for insights that can be translated into hypotheses to be tested. Drawing an analogy between a womb and a tomb portrays the woman/mother in a negative sense. Can we find supporting evidence for this negative image of the mother/woman through a comprehensive analysis of the sonnets? What can we say about a person who holds such a negative view of mothers/women? Can we find through this analysis indirect supporting evidence of Shakespeare's hypothesized homosexuality? These are interesting questions, and this chapter concludes not by providing answers but by pointing to a direction that will hopefully be examined by others.

References

Matte-Blanco, I. (2012). *The unconscious as infinite sets: An essay in bi-logic*. London, UK: Karnac Books.

Murtagh, F. (2014). Mathematical representations of Matte Blanco's bi-logic, based on metric space and ultrametric or hierarchical topology: Towards practical application. *Language and Psychoanalysis, 32*, 40–63.

Murtagh, F. (2016, in preparation). *Data science* foundations*: Geometry and topology of complex hierarchic systems and big data analytics*. Boca Raton, FL: Chapman\& Hall/CRC Press.

Neuman, Y. (2013). Shakespeare's first sonnet: Reading through repetitions. *Semiotica, 195*, 119–126.

Turney, P. D. (2013). Domain and function: A dual-space model of semantic relations and compositions. *arXiv preprint arXiv:1309.4035*.

Chapter 8
The Complexity of Personality: From Snowden to Superman

Up to now, I have criticized simple theoretical models of personality and the way in which the complexity of human personality has largely been ignored in personality research as well as in the field of computational personality. Let me try to describe personality by adopting a more complex, dynamic and one may even say radical approach.

My first argument is that personality is not a monolithic and static object but a heterogeneous dynamic system. Therefore, personality exhibits *several stable states* (e.g. attractors) rather than one. This means that, when studying human personality, we cannot aim to exhaust it by using a single or even a few fixed personality tags (e.g. extravert). Rather, personality as a dynamic system is like a landscape we tour and where we "rest" in various valleys. In addition, personality may be described as a "dissipative dynamic system" that is governed by *non-linear* dynamics. Non-linearity is easy to prove. What we describe as the dimensions of personality cannot be modeled as linear functions of other variables where a constant change in the independent variable is expressed as a constant change in the personality dimension. Think, for instance, about being open to experience as a personality dimension. Can you imagine a linear model that explains this variable by expressing a constant change as a response to a constant change in another variable? Is it possible to argue that openness to experience is a linear function of liberal education? Is it possible to argue that an obsessive personality, or what the old Freudians called an "anal" personality, is a function of tough parenting practices during toilet training, as they hypothesized?

The idea of personality as a dissipative system (or structure) is a little more challenging. A dissipative system is far from equilibrium and its structure is characterized by symmetry-breaking and complex structures. What has this physical terminology got to do with human personality and its study through computational tools? Think about the irreversibility of time. Our sense of self is to a large extent governed by an *autobiographical memory* through which we try to make sense of our experiences as individuals by weaving them into a coherent story. This effort is made because time is irreversible and because our self-concepts cannot be formed along a symmetrical axis. For example, as a young man I cannot truly see

© Springer International Publishing Switzerland 2016
Y. Neuman, *Computational Personality Analysis*,
DOI 10.1007/978-3-319-42460-6_8

myself as an old man by placing a symmetrical axis at the midpoint in my life, in the same way that, as an old man, I cannot truly see myself as a young man but only through the constrained spectacles of my present situation and past memories. Time flows in a reverse order only in fiction, such as in Philip K. Dick's story "Your Appointment Will Be Yesterday," and therefore symmetry-breaking and irreversibility are inevitable characteristics of our personality.

Here is another characteristic of personality as a complex system: personality is described by a large number of dynamic variables and thus has "high phase-space dimensionality." This means that it is improbable that we will be able to understand personality with just a few variables and a low-dimensional space. Tagging someone as a narcissist is a very important heuristic, but the complexity of personality is such that, even in the case of ultimate narcissists, this personality tag is of a quite limited value and we should strive to do better.

Finally, there is often *noise* in the dynamics, and parameter uncertainty in the available models of personality constructed from data. This uncertainty is not unique to human personality. When biologists are trying to predict the three-dimensional structure of a molecule by using its lower dimensional structure, they encounter the same problem. Determining the three-dimensional structure of a molecule—its "conformation"—is an extremely important challenge with far-reaching consequences. It is actually an attempt to predict the spatial arrangement of the atoms that comprise the molecule. The problem in predicting the three-dimensional structure of a molecule through its atoms and the way they are connected in two dimensions is that these small particles form a potentially infinite number of spatial arrangements. A recent breakthrough in the challenge of the "blind test of organic crystal structure prediction Methods" (Gibney 2015) suggests that quantum mechanical interactions must be taken into account to gain the best results. It means that, even to predict the three-dimensional structure of a molecule, we must to take into account the uncertainty that exists at the micro level of interactions. When we complain that human behavior is unpredictable even when a personality has been "diagnosed," I wonder whether we are in a similar situation to scientists who have just realized that understanding micro-level and "noisy" interactions is crucial for predicting the macro-level structure of a system.

Up to now, I have described personality as a complex system and pointed to its complex features. Now I can tell you that the above-mentioned characteristics of human personality are actually paraphrases of the characteristics used by Motter (2015) to describe real-world networks. I have used a rhetorical move in order to illustrate how easily and convincingly we may apply the language of complex systems to human personality. One must admit that these characteristics make a lot of sense when applied to human personality.

Let me delve deeper into the characteristics of personality as a complex system. As previously discussed, when studying personality in time, we do not recognize a fixed value (e.g. a person is not always an extravert) or something that can be meaningfully approximated by a linear function. This is what makes personality so complex. Human personality is far from equilibrious and always interacts with its environment. However, personality may be considered to be a dissipative structure—a kind of a steady state, like a hurricane. As argued by Harré (1998, p. 7):

> It seems to me that people produce streams of actions, some private, some public. These display all sorts of properties some of which we pick out as mental. There are stabilities and repetitions that recur in these streams of action, like vortices in a swiftly running stream.

So we have some kind of what Harré calls "unity of self," which is formed through the way we "tell" our personality. According to this perspective, human beings form narratives of their self, and through these narratives form their uniqueness and the unity of their self. This is why neuroscientists have emphasized the importance of autobiographical memory in the formation of the self.

The coherence of the self's narrative is important for mental health, as we know that *incoherence* of the self is an indication of a mental pathology. A deep feeling that your self is incoherent means that you cannot organize your emotions, experiences, memories and beliefs into one organized story and, from a psychological clinical perspective, this is a very undesirable situation. Therefore, analyzing the narrative of the self may be of high diagnostic value. For example, if the temporal order of events comprising a person's self-narrative is confused, this may be a sign of the fragmentation of the self. In this context, autobiographical memories are of the uttermost importance in forming the self and diagnosing personality.

Following Harré's concept of "unitas multiplex" (unity in diversity through purposiveness), in an attempt to understand a person through the texts that person creates, we may segment the texts into (1) the way the past is conceived through a form of autobiographical memory, (2) present actions and (3) future intentions. That is, in an attempt to understand personality as a complex and "constructed" system, we may focus on (1) autobiographical memories, (2) current actions and (3) future plans. Let's illustrate this idea by analyzing the famous letter written by Edward Snowden (see Appendix).

Snowden, according to some an admired and courageous whistleblower and according to others a traitor to be despised, has caused serious concerns for the US government after exposing the intrusive way in which the National Security Agency (NSA) and other intelligence agencies spy on citizens. Snowden is not a traitor motivated by greed but a person who explains his actions by giving them serious moral and ideological justifications, as is evident in his letter.

First let's try to gain a quick understanding of Snowden's letter by using IBM's natural language processing engine: Alchemy.[1] Alchemy identifies in the letter several entities such as the CIA and the National Security Agency, both of which are, surprisingly, judged by the engine as being loaded with positive sentiment. Overall, the document is analyzed by this engine as positively loaded and this is of course a problematic conclusion as the letter is highly negative with regard to several issues (e.g. it opens with the declaration: "I witnessed systemic violations of law by my government"). Although there are expressions of hope in the letter, a text

---

[1]http://www.alchemyapi.com/products/demo/alchemylanguage.

that exposes such a scandal as mass surveillance of citizens by a government cannot be described as positively loaded. In terms of emotions, the analysis provided by Alchemy is much better and exposes anger (0.59), sadness (0.14), disgust (0.15), joy (0.07) and fear (0.03). This analysis, though, is of minor informative value and seems to rely on surface words only. Reading this short letter, it is unlikely that one would judge "anger" to be the leading emotion. As we can see, a shallow automatic analysis of the letter is of minor relevance. At this point, let's try to dig deeper into the text by identifying the major arguments of Snowden and arranging them according to the narrative of past memories, present actions and future intentions.

> Past actions:
>
> 1a. I believe I witnessed systemic violations of law by my government that
> 1b. created a moral duty to act.
> 2a. I have faced a severe and sustained campaign of persecution that
> 2b. forced me from my family and home.

First, let us look closely these reflections on the past through an in-depth discourse analysis. Snowden opens with a testimony that he witnessed violations of law. The use of the hedge "believe" is highly informative. Snowden is sure that he witnessed violations of law. According to his letter, there is not even a slight doubt that he witnessed violations of law. He could have simply said, "I witnessed violations of law," but he chose to use the hedge "believe" as a form of rhetorical ambiguity despite the fact that he has no doubts about what he witnessed. A possible explanation and hypothesis is that Snowden uses the rhetorical language of the current post-modern Western intellectual elite, where a grain of doubt must be inserted even into grounded theoretical statements, empirical evidence and solid beliefs. Euclid, if living today and giving a talk at a distinguished university, could not have talked openly to students of the humanities and the social sciences by saying: "A point is that of which there is no part." He would have to rephrase his opening definition of the *Elements* by saying: "*I believe* that a point is that of which there is no part, *and a point is of course a subjective and culturally invented construct.*"

Recently, I gave a talk to graduate students of a science, technology and society program in one of the Israeli universities. Their reaction to the idea that personality has some logic beyond a relativistic, culturally constructed approach, and that we can apply computational tools to study personality, was a burst of infantile rage. One of the hot-tempered respondents criticized me for ignoring the fact that personality dimensions are subjectively and culturally "constructed" and accused me of using technology in the service of evil forces. As I'm very familiar with this dogmatic approach, I responded by using one of the cases presented in my talk, which was the automatic identification of pedophiles on the internet. I argued that I agree that pedophilia is a culturally constructed concept, and that I'm not so naive to argue that pedophilia is a form of social deviation in all possible worlds or even in all cultures and societies that have existed or still exist on this earth. The idea that children should be protected from being sexually abused by adults, if I may use a term loaded with value judgment such as "abuse," is definitely a culturally constructed concept. So what? Does this mean that pedophilia is legitimate and that we should not fight against it? Does it mean that human trafficking and the sexual abuse of children are things to be set aside until proven to be mathematically valid in all possible worlds? People who argue to the contrary are clear representatives of a certain cultural dogma that is no less dangerous than the dogma against which it rebels. Now, let's return to Snowden, the liberal whistle-blower who found shelter in one of the most democratic and liberal states on earth... Russia.

Snowden's use of the word "witnessed" is loaded with meaning. The verb "witnessed" belongs in the juridical context. A witness, usually an eyewitness, appears in court to describe what he saw, not to present his "beliefs." Witnessing a violation of law, Snowden experienced a "moral duty to act." In terms of personality analysis, this is a highly informative part of his narrative. Not every witness experiences a moral duty to act. Can you guess which personality type experiences a moral duty to act? If you have learned the lessons that I have taught so far, you may jump to your feet and shout, "The obsessive–compulsive personality is motivated by a moral duty to act"! I believe that you are right and that we may produce the hypothesis that this personality dimension is highly important in characterizing Snowden and his motivation.

If we adopt some general transformation rules presented in the context of natural language inference, we can summarize the narrative presented by Snowden so far as:

Snowden witnessed a violation of law that created a moral duty to act.

Based on the partial analysis so far, Snowden has scored high on the conscientiousness factor of the Big Five and well as on the obsessive–compulsive dimension discussed in the psychodynamic theories. He portrays himself as a highly moral individual who was forced to act when he witnessed injustice. Now, let's move on to an analysis of Snowden's actions in the present.

Present actions:

1. I am currently living in exile.
2. I am heartened by the response to my act of political expression.

Of these two statements, the first is clearly a *result* of past actions. This relation of causality is implicit in the statement and would have been difficult to identify through computational tools, which in identifying causality usually rely on cue words that explicitly mark this relation (e.g. "because").

Snowden, the moralist who felt obliged to act in the face of injustice, paid the price and was forced to live in exile. At this point, readers may feel a sense of deja vu—a feeling that they have encountered this general form of narrative in the past. This narrative actually falls into the genre of the superhero (i.e. Snowden), who protects the public from the supervillain (i.e. the CIA and the NSA) and sometimes has to suffer a painful blow (e.g. exile and persecution), only to return in a better future. Indeed, Snowden's painful exile is softened by the empathetic reaction he has received to his acts, and this leads him to a vision of future hopes and intentions.

Future hopes and intentions:

1. the United States will abandon this harmful behavior.
2. I will be able to cooperate in the responsible finding of fact regarding reports in the media.

What can we learn from this short narrative analysis? First, Snowden's narrative as expressed in his short letter is highly indicative of his personality. While personality may be a complex structure, analyzing a person's narrative, as proposed by Harré, may provide us with a simple heuristic that, on the one hand, pays tribute to human complexity and, on the other hand, structures human personality through narrative analysis. Although we have used in the abovementioned example a qualitative interpretative analysis, we may use *automatic narrative analysis* as part of computational personality analysis. Researchers have developed various conceptions of a narrative, with accompanying computational models (e.g. Ouyang and McKeown 2014). When using automatic narrative analysis in order to understand the personality of an individual, we must form a link between the narrative and the personality diagnosis. Currently, there is no tool that performs this task. One may use the tools of computational narrative analysis, but these tools only tell us a structured story about the plot or the characters and cannot provide us with workable insights about the personality that is telling the narrative. One creative way to address this challenge may be to identify some correspondence between the narratives told by individuals; film personae, as identified for instance by Bamman et al. (2013); and dimensions of personality. Let me explain this idea in a nutshell.

As we have learned from Harré, our personality may be conceived as the stories we tell about ourselves. In a modern Western culture governed by mass media and the entertainment industry, some of these stories are framed through cinema or TV characters with which we identify. Similarly to the way we speak the language of music (Neuman et al. 2016), we speak the language of films. This is an instance of "intertextuality" that may be highly important in computational personality analysis. When analyzing Snowden's letter, I couldn't resist the feeling that this highly moral and obsessive personality describes himself through the fictive persona of Superman. Like Superman, who in his daily life is an unnoticed nerd, Snowden, whose appearance is not exactly conventionally heroic, emerges from his letter as a superhero who fights the dark forces of a powerful organization. Fusing the analysis of his narrative with the structure of film personae and personality dimensions may support this hypothesis and provide workable insights.

Regardless of the value judgment we may attribute to Snowden, conceived by some as a hero and by others as a villain, for intelligence agencies it may be highly important to identify the personality of what is called the "insider threat." Many years ago, the CIA made an attempt to profile the personality of spies in a project known as "SLAMMER." One of this project's reports recognized that a single profile doesn't exist but identified some personality traits—such as being antisocial and narcissism—shared by people who spied in the United States. Snowden doesn't fall into this simplified scheme but he does seem to be motivated by a heroic sense of self combined with highly moralistic and obsessive–compulsive personality dimensions. Recognizing and understanding this combination may be important for organizations such as those who hired Snowden and for other organizations, which may value having such productive and loyal workers until a sudden and unexpected catastrophe turns them into an "insider threat."[2] On the other hand, the personality dimensions of Snowden and others who follow his archetype may be desirable to those who would like to challenge organizations such as those who hired Snowden.

In this chapter, I have introduced the idea that, despite the complexity of human personality, there are some heuristics we may apply in order to better understand it, such as personality-guided narrative analysis. Finding structures in the dynamic nature of human personality is also the major theme of the next chapter, where I illustrate methodologies we may use to identify "attractors" of personality.

References

Bamman, D., O'Connor, B., & Smith, N. A. 2013. Learning latent personas of film characters. In *Proceedings of the 51st Annual Meeting of the Association for Computational Linguistics* (Vol. 1: *Long Papers*), pp. 352–361. Stroudsburg, PA: Association for Computational Linguistics.

Gibney, E. (2015). Software predicts slew of fiendish crystal structures. *Nature, 527*(7576), 20–21.

Harré, R. (1998). *The singular self: An* introduction *to the psychology of personhood*. London, UK: Sage.

Motter, A. E. (2015). Networkcontrology. *Chaos: An Interdisciplinary Journal of Nonlinear Science, 25*(9), 097621.

Neuman, Y., Assaf, D., Cohen, Y., & Knoll, J. L. (2016). Metonymy and mass murder: Diagnosing splitting through automatic text analysis. In M. Arntfield & M. Danesi (Eds.), *The criminal humanities: An introduction* (pp. 61–73). New York, NY: Peter Lang.

Ouyang, J. & McKeown, K. (2014, May). Towards automatic detection of narrative structure. http://www.lrec-conf.org/proceedings/lrec2014/pdf/1154_Paper.pdf. Accessed May 2, 2016.

---

[2]A wonderful example of this dynamic appears in Le Carré's novel *A Perfect Spy* (1986).

Chapter 9
Attractors of Personality: A Murderer, a Terrorist and Some Angry Jews

A dynamic system may visit various loci of its "phase space," the space where all of the system's possible states are represented. Regularity in the behavior of such a system can be found by identifying its "attractors." An attractor is a limited region of this space, where the system is "caught," like water running into a sink. For example, my house has at least four entry points. One of our cats uses all of these entry points but his trajectories are such that we always find him napping on his favorite couch. The couch is an attractor for the cat's spatial movement in the phase space comprising all of its possible napping locations.

As you may recall, when discussing human personality through the psychodynamic approach, I described it as involving themes or zones of preoccupation. These themes can be considered as attractors. For instance, you may recall certain people who, regardless of a conversation's starting point, always find their way to a certain theme, as if their mind is attracted to a certain region. I've a dear friend who is a neo-Marxist anthropologist of Argentinean origin. It has always struck me that, regardless of our conversation's starting point (whether literature, food or something else), I always find myself listening to his passionate attack against capitalism and the nation state. Another dear friend of mine has what you may call a narcissistic attractor. Regardless of our conversation's starting point, it always reaches a point where his self-praise is explicitly stated or implicitly implied. I jokingly explain to him that he is like the Greek King Midas, who turned everything he touched into precious gold, but also warn him that he should remember the tragic end of this mythological king.

Finding the attractors of personality is a challenge that can be addressed through some interesting search algorithms such as the "deterministic tourist walk." The deterministic tourist walk is simple to explain (Lima et al. 2001). Let's assume that we have a network (or a graph) consisting of nodes (vertices) and connections between these nodes (edges). At each time step, we move from node "x" to the nearest node that hasn't been visited in the previous m steps. This walk has a transient part that indicates the length of the chain we have passed until being caught in a cycle (i.e. an attractor) of a certain length "t." The memory window of the walk may be narrow, such as in the case that we move from node x_i to node x_{i+1}

© Springer International Publishing Switzerland 2016

Y. Neuman, *Computational Personality Analysis*,

DOI 10.1007/978-3-319-42460-6_9

unless we have already visited node x_{i+1} in the previous two steps. In this case the memory window "m" is equal to 2. It is argued that, when the memory window is low, the tourist walk reveals local features of the network and, when the memory window is high, global features are exposed (Silva and Zhao 2015).

As the deterministic tourist walk has been found to be simple and effective in identifying patterns (e.g. Backes et al. 2010), albeit in a totally different context from the one we discuss in this book, it is worth trying to use it to identify attractors of personality. To address this challenge, I present three case studies.[1] The first involves analyzing the "manifesto" written by Seung-Hui Cho, the mass murderer mentioned in Chap. 4 when discussing screening for school shooters. The second case involves a speech given by the arch-terrorist Osama bin Laden titled "Letter to the American People" (2002). The third case involves the analysis of 1500 blogs written by Jewish anti-Israeli activists. In the third case study, we don't analyze the personality of a single individual but the "collective self" of a group. Can we find interesting attractors in these texts by using the deterministic tourist walk? Can we interpret these attractors as personality themes? Let's start with Cho's manifesto. Note that the methodology (described in the next section) used to find attractors in Cho's manifesto was applied to the two other case studies as well.

9.1   Into the Mind of a Murderer

First, my colleagues and I used a dependency parser and represented the text as pairs of words associated via a syntactic relation. For instance, when Cho writes:

> You forced me into a corner and gave me only one option.

this utterance can be represented using the Stanford Parser, as follows:

```
root ( ROOT-0 , forced-2 )
nsubj ( forced-2 , You-1 )
nsubj ( gave-8 , You-1 ) (extra)
dobj ( forced-2 , me-3 )
case ( corner-6 , into-4 )
det ( corner-6 , a-5 )
nmod:into ( forced-2 , corner-6 )
cc ( forced-2 , and-7 )
conj:and ( forced-2 , gave-8 )
iobj ( gave-8 , me-9 )
advmod ( option-12 , only-10 )
nummod ( option-12 , one-11 )
dobj ( gave-8 , option-12 )
```

---

[1] I'm grateful to my programmer, Yochai Cohen, for writing the code for this analysis.

Next, we translated this dependency representation into a semantic graph that included only words that belong to four categories of language: nouns, verbs, adjectives and adverbs. Through "lemmatization," we grouped nodes of plural and singular form (e.g. "group" and "groups") and applied the deterministic tourist walk as follows. We chose each of the nodes/words comprising the graph and at each time step moved to the closest node by measuring the semantic similarity of the associated nodes using the similarity matrix developed by Turney et al. (2011), ensuring that the semantically closest node hadn't been visited in the previous m steps, as defined above. For each word and for each memory window, we identified the relevant attractor and the words from which it is comprised. Now, let's see what can we learn from this analysis.

Cho's manifesto includes 445 unique words. In order to understand whether there are commonalities shared by all of these words, we again used the natural language processing program Alchemy and found that the words used by Cho can be organized into three main taxonomies (with the lowest categories in bold):

1. crime/personal offense/**torture**
2. society/**sex**
3. crime/personal offense/**homicide**.

In addition, the two main organizing concepts of the manifesto are "fuck" and "hedonism." Therefore, a simple analysis reveals to us that this text has a negative sentiment and has something to do with violence and sex. When analyzing Cho's manifesto through the deterministic tourist walk, using a memory window of 1 and counting the words comprising the attractors reached from each and every starting point (i.e. word), we find the following words in descending order:

weak, poor, rape, innocent, give, feel, take, say, Jesus Christ, fucking, fuck, kill, hunt, steal

When these attractor words are inserted into the Alchemy engine, we find that they are organized under three main taxonomies: "torture" and "homicide" (under "crime") and "pagan" and "wiccan" (under "religion"). The most dominant concept in this attractor is "rape"; the concept dominating the chain of words leading to the attractor is "fuck"; and overall the attractor has a negative sentiment, as we might expect.

The added value of moving from analyzing all of the words used in this manifesto to analyzing the words that populate the attractors is the emergence of a religions pagan aspect of Cho's mind, which was hidden when we analyzed all of the words. We learn that, regardless of his starting point, Cho's mind is attracted to the lethal combination of violence and paganism associated with the word "Satan." As we may learn, the violence expressed in Cho's manifesto is not the "trivial" violence of crime and war that we meet in the news on a daily basis. It is violence with a very troubling religious aspect. In some respects, and for reasons not to be detailed here, a criminal who uses the language of religion is much more dangerous than a criminal who uses the typical language of street violence.

As we have explained, when the memory window is low, local features of the graph, which are much more salient, are exposed. However, from a psychological point of view, it may be much more interesting to expose long-range correlations that are less obvious in the text. To find these deep patterns, we should increase the memory size of our search.

When we increase m in our search to 2, we find that the words comprising the attractor can be organized by the automatic taxonomy analysis under the heading of "present-day saints," and when we increase m to 3 we find another taxonomic dimension of "death penalty." When we reach the maximum memory size of 4 and 5, which is the "complexity saturation" point (Silva and Zhao 2015) of our analysis, we encounter three main taxonomic dimensions: "death penalty," "children" and "disease."

What is the personality narrative that we may build based on this very simple analysis? It is primarily about a person who considers himself a victim of personal sexual offense (rape?) and torture. It is a person who is conceptualizing his pain in religious terms as a struggle between evil (i.e. Satan) and good (i.e. saints) and on the macro level is talking about the death penalty, probably in relation to those who have hurt him so much. This narrative can be further enriched by using other texts written by Cho and by using other measures we have developed for personality analysis. For instance, Cho conceptualizes his pain in binary terms as a fight between Satan and God. This is a clear indication of using the "splitting" defense mechanism, which was discussed in Chap. 2.

To recall, splitting is a very primitive defense mechanism that involves the compartmentalization of experience into "all-good" and "all-bad" categories with no room for ambiguity. People are either saints or Satan. Splitting is used to reduce anxiety but at the price of distorting reality (i.e. people are usually not all good or all bad) and at the price of chaotic shifts in feelings of self-worth and self-representations. In a previous work, which aimed to automatically identifying splitting in texts (Neuman et al. 2016), my colleagues and I used two simple heuristics. First, we examined the emotional valence of words associated with the self and others through the analysis of words attached to the first-person pronoun "I" and to the pronoun "you" in the texts of three mass murderers, including Cho. Second, we measured the level of *ambiguity* in the texts and, as we hypothesized, individuals who see the world in terms of black and white don't have any space for healthy ambiguity and uncertainty. Again as hypothesized, we found that Cho's manifesto scored high in terms of polarity of valence within the self (i.e. indicating chaotic shifts in self-representation) and between the self and others, and low in terms of ambiguity, which indicates a text where there is no room for doubt.

These results converge with the attractors, or areas of preoccupation, which we identified before. An interesting point concerns Cho's preoccupation with victimhood resulting from sexual abuse and with bloody revenge or the death penalty, as are evident on the deepest level of the graph. Cho was also the author of a repulsive drama script titled "Richard McBeef," featuring a child who accuses his stepfather of pedophilia and finally murders him; thus, in Cho's manifesto, the death penalty and its association with violent sexual abuse and rape echo the same themes in an earlier text written by the same murderer. From a psychological point of view, the convergence of these themes cannot be dismissed as a mere coincidence. We don't know whether in his childhood Cho was a victim of sexual abuse. However, our computational tools powerfully suggest this hypothesis and may explain his violent outburst as a delayed response to such a trauma. If this is the case, the victims of his massacre were the unfortunate targets of his violent projections resulting from wrong experienced in his childhood.

In sum, identifying the attractors in Cho's manifesto has given us interesting insights into his personality as a complex and dynamic structure. The analysis does not provide us with a psychological interpretation but does excavate hidden patterns that are highly important for an in-depth personality analysis. The next section introduces a similar analysis, this time of a well-known terrorist.

9.2 Into the Mind of a Terrorist

The next text used to illustrate attractor-based analysis is a speech given by Osama bin Laden, the arch-terrorist and the mastermind behind the terror attacks of September 11. Bin Laden's speech has 794 unique words, which, when organized into a taxonomy, teach us that the speech is about three major themes: Islam, Judaism and "unrest and war." The dominant concepts evident among these words are those of "humiliation" and "oppression" and overall the speech has a negative sentiment.

When we use the deterministic tourist walk with $m = 1$ and analyze the words comprising the attractors, we identify the themes of "unrest and war," "racism" and "government." When we move to $m = 2$, we identify (similarly to in the case of Cho) the themes of "personal offense" and "humiliation." When we move to $m = 3$, we identify the themes of "latter-day saints" and "government" and the overall sentiment is positive. The same themes remain constant for $m = 4$ to $m = 5$. However, when we reach the final attractors at $m = 6$ and $m = 7$, we find again the theme of "unrest and war."

The surface level of the text deals with trivial and expected social and political issues. However, when we move deeper into the text through the tourist walk, we find the theme of *personal* offense. In this case, building a narrative out of the emerging attractors seems to be little bit more complicated than in the case of Cho. What is the narrative that pops out of bin Laden's speech?

On the surface level, bin Laden's speech is about political issues per se and as such seems to be of little informative value for learning something about the personality of this vicious terrorist. However, when we dig deeper into the speech, we find that it is not only about abstract ideological or political issues but about *personal* offense and humiliation that is translated into the ideal of war in the name of religion, as is evident from the positive sentiment associated with the deeper layers of the speech. In other words, the "manifesto" of this arch-terrorist teaches us something about the terrorist mind, something that is shared by many—too many—of the terrorists operating in the name of Islam. The story is about the *personal humiliation* resulting from a cultural–political clash. This is an important topic and touches the most delicate aspects of cultural psychology.

The terrorists who launched the deadly terrorist attacks in Paris (November 2015) in the name of ISIL followed the same language of personal humiliation. Some of the terrorists were Islamic European citizens who had experienced their meeting point with their host society as humiliating. French society is composed of various groups of immigrants from various states (e.g. Spain), regions (Asia) and religions. Most of these groups have been somehow assimilated into French society without experiencing their situation as a personal humiliation. Bin Laden's speech gives us a glance into the mind of a terrorist that represents the minds of others. Those who experience deep personal humiliation as a result of a situation that is not necessarily personal (i.e. bin Laden was not personally humiliated by the United States before launching his attacks on September 11) are personalities we may describe as disturbed narcissists.

As you may recall from Chap. 1, a narcissist is someone who shows signs of vulnerability underlying an inflated facade of self-importance. Narcissists have a deep problem with self-esteem, and when feeling inferior they tend to be angry and competitive and often respond to what they conceive as a degrading experience with a narcissistic rage sometimes expressed as bloody vigilance. Analyzing the attractors of bin Laden's speech, we may hypothesize that he was a disturbed narcissist who experienced the cultural encounter with the West as humiliating and derogatory, such that it could be resolved only through projected anger. These emerging themes and the associated diagnosis may be extremely helpful when trying to screen for potential terrorists, who may have similar states of mind to bin Laden's. Again, it must be emphasized that the identification of attractors through computational methods may provide us with a window into the complexity of personality. However, a window cannot replace the observer, who should look through it and make sense of what he sees. This approach is clearly in line with the idea of "augmented intelligence" rather than with what can be the ungrounded pretentions of artificial intelligence. The next case study provides additional evidence of the way in which identifying the attractors of personality may enrich our understanding.

9.3 Into the Minds of Self-hating Jews

An act of voluntary religious conversion isn't only an act of faith but also a process that involves a problem with one's personal, psychological and social identity (Leone 2004). Moreover, it is a process of *destabilizing* the identity and unity of the self (Leone 2004) and therefore may be of great interest to psychologists who are interested in personality and its transformation.

Why should someone be interested in the psychology of conversion? While religion in some ways seems to be a marginal force in modern Western societies, the rise of what is called political Islam has brought it back into the forefront in two main senses. First, the rising force of radical Islam strives to convert non-believers and members of other religions. For instance, the Sharia for UK movement and its European replications have as their ideal the conversion of Europe to Islam. The second sense in which conversion is evident in the current discourse concerns the conversion of young people to Islam as a part of a radicalization process. This voluntary conversion involves Muslims and non-Muslims alike. In this context, it is of great interest to study the way in which radicalization, specifically violent radicalization, is entangled with religious conversion and the transformation of the self. Here, I don't deal with this Islamic aspect of conversion but with another issue.

While up to now conversion has been described in the psychological literature in psychological–individualistic terms, conversion is an act of identity transformation in which one's *collective identity* is *proactively* rejected in favor of another collective identity. Rejection, at least from a psychological perspective, seems to be an important aspect of conversion. As rejection is loaded with a strong negative valence, we may expect it to be accompanied by *self-hate*. After all, rejecting your former identity is rejecting a constituting part of your *own self*, and hence your *personality*. This dynamic explains why in European languages "tradition" and "treason" share a common etymological root, and why treason is conceived as a violation of cultural transmission/tradition (Leone 2004).

An interesting case epitomizing the dynamics of conversion, destabilization of the self and self-hate is that of the philosopher Otto Weininger. Weininger was born to an assimilated Jewish family in Austria in 1880. Like many other assimilated Jews of his period, he voluntarily rejected his Jewish identity and converted to Christianity in order to avoid the anti-Semitism of the Austrian society and to gain its acceptance. However, the Austrian society did not warmly accept the assimilated Jews.

Weininger, held to be a prodigy, completed his Ph.D. in philosophy at a very young age but was a tormented person who struggled with two main conflicts: being a Jew in a Christian, anti-Semitic society and being a latent homosexual. In 1903, at the age of 23, he committed suicide after writing his famous work *Sex and Character* (Weininger 1903/2005), which was intensively used by the Nazis. He was a person who struggled with his Jewish as well with his homosexual "feminine" identity, so it is not surprising to find that the book is a polemic treatise (illustrating the relation between conversion and controversy; Leone 2004) that praises the masculine character and makes a pseudoscientific attempt to establish the inferiority of women and Jews. However, in this manifesto of self-hate, we find a rare moment of reflection where Weininger says: "we hate in the other what we are but afraid to be." This is our point of departure; let's show how we can expand on it by analyzing the case study of anti-Israeli Jewish organizations.

Although Israel has been the promised land of the Jewish people for thousands of years, and its capital—Jerusalem—has never been the capital of another nation, the establishment of the State of Israel at the late 1940s has been negatively perceived by two main Jewish sectors. The first is the ultra-orthodox sector, which opposed the proactive nationalist and secular Jewish approach and considered divine intervention as the only legitimate reason for the establishment of a Jewish state. The second sector comprises assimilated Jews who considered the "aggressive" and proactive Israeli Jewish identity as a barrier to an adaptive and silent assimilation into modern Western states. While the ultra-orthodox group is to a large extent indifferent to the State of Israel, the second group, which is mainly composed of non-orthodox secular Jews, has been proactive in its opposition. For these Jews, the Jewish Israeli identity can be perceived to reflect what they are "afraid to be." In this context, the activities of Jewish anti-Zionist organizations are conceived by many Jewish Israelis as acts of treason reflecting self-hate, which has been a persistent theme among modern western Jews (see Gilman 1986) since the European emancipation of Jews. Therefore, studying the collective personality of Jewish anti-Israeli organizations through computational tools is an interesting case study in better understanding collective personality, identity and a modern form of conversion. Can we apply our attractor identification methodology to gain insights into the "collective personality" of anti-Israeli Jews?

Our data include texts written on the blogs of several major anti-Israeli organizations. We start with the blog "It's Kosher to Boycott Israeli Goods," which is the official blog of the organization Jews for Boycotting Israeli Goods (J-BIG). J-BIG—is a group of Jews residents in the UK who came together in 2007 to support the Boycott Israeli Goods campaign. We identified 136 texts from this blog. Through this blog, we identified two additional Jewish anti-Israeli blogs. Jews Sans Frontiers is an independent anti-Zionist blog. Its author scans the web and gives updates on recent news and propaganda. From this blog we drew 1310 texts. Jews Say No! is a group of "concerned Jews" in and around New York City that formed in winter 2008–2009 to express their opposition to Israel. Interestingly, they describe themselves as participating in demonstrations holding signs such as "Am I a self-hating Jew if I oppose illegal and inhumane policies of the Israeli government?" From this blog we drew 62 texts. Overall, 1508 texts were analyzed.

As previously explained, each text was parsed using the Stanford Parser and a semantic graph was built of all the texts by using the lemmas of nouns, adverbs, verbs and adjectives. The general graph was composed of 1171 nodes (i.e. words) and 2163 edges (i.e. dependency properties).

First, we used attractor analysis through the deterministic tourist walk algorithm, as described previously. When analyzing the words constituting the attractor of the longest memory window, we find that these words are organized under the taxonomy of (1) racism, (2) human rights and (3) unrest and war. The main *concepts* organizing this attractor are (1) racism, (2) xenophobia, (3) nationalism, (4) discrimination, (5) race, (6) Nazism, (7) prison and (8) fascism. These concepts give us a quick glance into the collective mind of the blogs' writers. If the positive identity of the anti-Zionist Jew is actively formed against the "non-self" of the Israeli Jew, it is formed as the mind of a modern liberal Jewish thinker against what is conceived as a nationalist, racist, fascist, Nazi, xenophobic and discriminatory identity of the Israeli Jew.

This interpretation is in line with Cuddihy (cited in Gilman 1986, p. 15), who argues that the conflicts within modern Jewish identity are the result of the

Enlightenment ideal of civilized behavior and the attempt to reject "uncivilized" behavior in favor of "civilized" behavior. The "uncivilized" behavior, argues Cuddihy, is associated with the Eastern Jew, and the civilized behavior is associated with Western enlightened society. While the "civilized" object (i.e. Westerners) has remained constant through the years, it seems that the uncivilized object, which in the past was associated with the "savage," is now associated with the Israeli Jew. This interpretation exposes in our data an interesting triadic structure in which Jews are contrasted with Israelis and associate coalition with the "oppressed" Palestinian group. This dynamic is in line with liberal ideas of recognizing and empathizing with the rights of oppressed minorities. To test this hypothesis, we identified all words associated in the graph with the lemmas of the words "Israelis", "Palestinians" and "Jews", and analyzed them through the LIWC. From now on, using the terms "Israelis", "Palestinians" and "Jews", is in the sense of the signs as they appear and represented in our corpus and not in the sense of the actual groups.

When we use the LIWC's categorization, several interesting findings pop up. The first concerns the use of pronouns. The use of the personal pronouns "I" and "we" is more salient for Jews than for Palestinians or Israelis. Therefore, the texts expose a general notion of collective identity in which the authors affiliate themselves with some kind of Jewish identity, whatever that may be. However, and this is the important point, the language of "I" and "we" is much more pronounced in the case of the Palestinians than in the case of the Israelis. This finding may indicate that a boundary has been formed between the Jewish identity and the Israeli identity, as we hypothesized from the analysis of the attractors. The authors of the blogs may want to emphasize that, while they are Jews, they are *not* Israeli Jews, and that their heart as emancipated, enlightened and liberal Jews is with the Palestinians. This conclusion is strengthened when we find that the "you" pronoun is associated more with the Israeli than with the Palestinians.

If "we" are Jews and our collective identity is extended to the Palestinians and contradicted by the one of the Israelis, what can we say about our new "converted" collective self which is contrasted with the "non-self" of the Israeli Jew? Jews are associated with *sad* words, similarly to the Palestinians, while Israelis are more associated with *anger* words. Interestingly, Palestinians are more associated with body and sexual words, a finding that supports the interpretation that the blogs of emancipated Jewish writers hold an "orientalist" representation of the "noble savage" with all of its accompanying sexual connotations. Jews are also associated with more present and future terms while Israelis talked more about in terms of past tense. Israelis were the most associated with words to do with money, religion and swearing. In comparing Jews to Muslims, for instance, we noticed that Jews scored higher on money, swearing and power words.

If there is a lesson to learn from this simple analysis of words, it is that the writers of the blogs identify themselves with Jews and differentiate themselves from Israelis. However, what the Jews and Israelis seem to share, in terms of their representation in the semantic graph, is their dominance (power) and money, which are major themes of anti-Semitic thinking. These findings raise the hypothesis that, when the Jewish authors are involved in anti-Zionist activities, they engage in self-criticism, which is known to be associated with depression.

Westen et al. (2012) generated an empirically derived personality taxonomy in which personality types are organized according to two main clusters, or what they describe as "spectrums." The "internalizing spectrum" involves individuals who on the extreme end of the spectrum might experience painful emotions, mainly depression and anxiety. They are emotionally inhibited and socially avoidant and tend to *blame themselves* for their suffering, therefore turning their rage against themselves. This spectrum includes the following personality types: depressive, anxious–avoidant, dependent–victimized and schizoid–schizotypal. In contrast, those who reside on the "externalizing spectrum" turn their anger and aggression against others. This spectrum includes the paranoid and the narcissist personality types. A major difference between the two personality axes is the way in which people *manage their aggression*. This taxonomy invites the hypothesis that the anti-Zionist Jews will be characterized by higher level of depression when writing about Jews and by higher scores on the externalizing spectrum when talking about their Israeli "non-self." The rationale for this hypothesis is as follows. For many years, European Jews lived in a hostile, humiliating and aggressive environment in which they were unable to express their aggressiveness. This form of socialization created, together with other complex cultural threads, a form of self-criticism in which the aggressor point of view is internalized as a form of defense mechanism. In this context, the Israeli approach, which is clearly an aggressive approach, may pose a threat to the assimilated Jewish identity. An assimilated Jew is a "hidden Jew" who strives to ensure that the Jewish identity will not be imposed on him or at least will remain unnoticed.

To test the above hypothesis, we have used our vectorial semantics approach to personality analysis and measured the similarity between the personality vectors used in Neuman and Cohen (2014) and the vectors of the words associated to the first degree with the lemmas of Israelis, Jews and Palestinians. We first hypothesized that Jews would score higher on average than Israelis on the dimensions of the internalizing spectrum and Israelis would score higher on average than Jews on the dimensions of the externalizing spectrum. Indeed, Jews scored slightly higher on the internalizing spectrum (0.09 vs. 0.07 respectively) and Israelis higher on the externalizing spectrum (0.29 vs. 0.09 respectively). To test this hypothesis from another perspective, we used the externalizing and introvert vectors (Neuman and Cohen 2014) and measured their similarity to the Jews' and Israelis' vectors. In this case, too, Jews scored higher on the introvert measure (0.10 vs. 0.04 respectively) while Israelis scored higher on the extravert vector (0.22 vs. 0.09 respectively). More specifically, I have hypothesized that Jews will score higher on the depressed dimension while the "aggressive" and "arrogant" Israelis will score higher on two personality dimensions of the externalizing vectors—paranoid and narcissistic—corresponding to their aggressive and arrogant representation in the blogs. Table 9.1 presents the results.

As we can see, and in a sharp contrast with my own expectations, the results clearly support my hypothesis and provide some evidence that self-hate among

Table 9.1 Jews' and Israelis' similarity in terms of the personality vectors

	Jews	Israelis
Depressed	0.09	0.03
Narcissistic	0.09	0.19
Paranoid	0.09	0.20

Jews exist as a part of identity conversion. In sum, in this chapter, I have attempted to inject more complexity into computational personality analysis by showing how attractors of personality can be identified. These attractors may be directly interpreted or may serve as an input for another form of analysis. Defining the data in which we identify the attractors and the exact methodology in which the attractors are identified should be a reasoned process led by the researcher in order to fit the goal of the project. If there is one attractor that should be avoided it is the attractor of dogmatic thinking…

While, up to now, I've attempted to apply some basic ideas of complex systems to the analysis of personality, these ideas were quite conservative. The following chapter takes us a step forward in the computational study of personality as a complex system.

References

Backes, A. R., Gonçalves, W. N., Martinez, A. S., & Bruno, O. M. (2010). Texture analysis and classification using deterministic tourist walk. *Pattern Recognition, 43*(3), 685–694.

Gilman, S. L. (1986). *Jewish self-hatred*. Baltimore, MD: Johns Hopkins University Press.

Leone, M. (2004). *Religious conversion and identity*. London, UK: Routledge.

Lima, G. F., Martinez, A. S., & Kinouchi, O. (2001). Deterministic walks in random media. *Physical Review Letters* 87(1), 010603.

Neuman, Y., & Cohen, Y. (2014). A vectorial semantics approach to personality assessment. *Scientific Reports, 4*, 4761.

Neuman, Y., Assaf, D., Cohen, Y., & Knoll, J. L. (2016). Metonymy and mass murder: Diagnosing splitting through automatic text analysis. In M. Arntfield & M. Danesi (Eds.), *The criminal humanities: An introduction* (pp. 61–73). New York, NY: Peter Lang.

Silva, T. C., & Zhao, L. (2015). High-level pattern-based classification via tourist walks in networks. *Information Sciences, 294*, 109–126.

Turney, P. D., Neuman, Y., Assaf, D., & Cohen, Y. (2011). *Literal and metaphorical sense identification through concrete and abstract context*. In *Proceedings of the 2011 Conference on Empirical Methods in Natural Language Processing, Edinburgh, Scotland, UK, July 27–31* (pp. 680–690). Stroudsburg, PA: Association for Computational Linguistics.

Weininger, O. (1903/2005). Sex and character: An investigation of fundamental principles. Bloomington, IN: Indiana University Press.

Westen, D., Shedler, J., Bradley, B., & DeFife, J. A. (2012). An empirically derived taxonomy for personality diagnosis: Bridging science and practice in conceptualizing personality. *American Journal of Psychiatry, 169*, 273–284.

Chapter 10
Taking Complexity a Step Forward: The Reversibility of the Pedophile's Mind

10.1 Reversibility and Irreversibility

Previously, I have pointed out that personality, as a complex system, is irreversible, as the time course of its manifestation cannot be inverted or reconstructed. However, the idea of reversibility has challenged the minds of philosophers and authors alike. In *The Gay Science*, Nietzsche (1882/2010) writes:

> Your whole life, like a sandglass, will always be reversed and will ever run out again.

This idea suggests that reversibility is possible as life recurs in repeating patterns, but it is clear to any reasonable person that, beyond deep philosophies and imaginative fictions, the sandglass cannot be reversed. However, recognizing a certain degree of reversibility may be important in understanding the behavior of a complex system.

A process is defined as time reversible (Weiss 1975), statistically speaking, if for every N the series $\{X(t_1),\ldots, X(t_N)\}$ and $\{X(t_N),\ldots, X(t_1)\}$ have the same probability distribution. It has been shown that in certain biological systems irreversibility decreases with aging and pathology, which means that a certain level of irreversibility is a tool for maintain flexibility, health and resilience. This idea seems to contradict the intuition of those interested in fiction involving time reversibility; being able to go back in time (i.e. reversibility) seems to be necessary to correct the past and change the present and the future. This idea doesn't have to involve only science fiction. Our ability as intelligent creatures to form our past in a constructive and creative sense is a clear indication that some form of reversibility exists in the realm of psychology and is important for maintaining a healthy personality.

In some psychology classes, I give my students the exercise of writing their autobiography in both a prospective (i.e. as a child seeing his life unfolding) and a retrospective (i.e. from the perspective of the adult) manner, and ask them to compare the two versions of their stories. However, being able to reverse time, even as an exercise in imagination, cannot dismiss the idea that a living functional system

© Springer International Publishing Switzerland 2016
Y. Neuman, *Computational Personality Analysis*,
DOI 10.1007/978-3-319-42460-6_10

is basically irreversible. For example, let's assume that we measure the sentiment expressed by a certain individual on a scale ranging from very positive to very negative. The measurement takes place in a specific context, in a well-defined time period and so on, and should be indicative of the individual's personality. A person who is fully "reversible" is a person whose emotional expression is predictable and hence rigid. A person whose time series of emotion is more irreversible is a person who expresses a higher level of "freedom" and therefore adaptability. In sum, we may hypothesize that measuring irreversibility/reversibility of time series that express certain personality dimensions may be indicative of the individual's personality, but is it possible to measure reversibility and irreversibility?

Let me give you just one example of a methodology for addressing this challenge (Lacasa et al. 2012). We first map our time series into a network as follows. Let's suppose that we have a time series of 1, 2, 1, 2. Two nodes (i and j) in the graph are connected if one can draw a horizontal line in the time series joining xi and xj that does not intersect any intermediate data height. See Fig. 10.1, where our abovementioned time series is represented. This representation is then mapped as shown in Fig. 10.2.

We can see that node 1 receives no arrows and sends one arrow, node 2 receives one arrow and sends two arrows, and so on. Therefore, we have in-degree and out-degree distributions on our graph. The irreversibility of the time series is measured as follows, using the Kullback–Leibler divergence measure:

$$D[Pout(k)||Pin(k)] = \sum_k Pout(k)log\frac{Pout(k)}{Pin(k)}$$

To illustrate this measure, I used a corpus of conversations provided to the participants of the International Sexual Predator Identification Competition at PAN-2012 (Inches and Crestani 2012). From this corpus, we randomly selected one conversation between a pedophile and his "victim," usually a volunteer aiming to

Fig. 10.1 A visibility graph of the time series

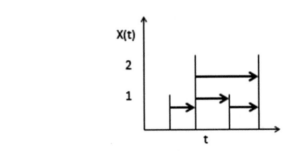

Fig. 10.2 A representation of the visibility graph

catch the pedophile. We hypothesized that the pedophile is obsessed by lust and that his conversation is pathologically organized around this theme and the attempt to seduce his victim. Therefore, it can be expected that the appearance of lust in his utterances will be more reversible than the appearance of lust in the victim's utterances as the victim is not motivated by the same pathology. We translated the words used in each exchange of this conversation into vectors of words (one for the pedophile and one for the victim) and measured their distance from a vector of lust words. The numbers produced through this procedure is therefore a measure of the extent to which the words used by either the pedophile or his victim express lust at each time point of the conversation. When analyzing that data and comparing the time series,[1] we found that the time series of the victim was significantly more irreversible than the one produced by the pedophile (0.10 vs. 0.02 respectively, while the irreversibility score of the null model was 0.07 with a standard deviation of 0.028). It goes without saying that this analysis of irreversibility and its results are for illustration only. However, the idea that the time series of personality dimensions may be analyzed and contribute to computational personality analysis is an important point. Analyzing the dynamics of personality is also the focus of the next section.

10.2 Recurrence Quantification Analysis

In the previous section, I showed how analyzing a time series of personality data may help us to better understand personality through its temporal aspect. Another possible direction is to use recurrence quantification analysis (RQA). Here, I briefly explain RQA and illustrate how it may be applied to the analysis of personality.

When we measure a certain personality dimension over time, we may be interested in identifying the unique temporal signature of its appearance—for instance, how "chaotic" or "deterministic" it is. The methodology of RQA (Webber and Zbilut 2005) is a relatively new approach that aims to identify and quantify different signatures of repetitions. The idea of repetition is highly important in identifying a system's temporal signature, from the analysis of heartbeats to the analysis of human personality. I have previously shown how repetitions are evident in Shakespeare's sonnets and how a careful analysis of these repetitions may serve as a gate into Shakespeare's mind. Here I approach the analysis of repetition in a different way and through the methodology of RQA. An excellent introduction, which is accessible even to the non-expert, appears in Webber and Zbilut (2005), and here I provide a very brief introduction.

---

[1]I'm grateful to Lucas Lacasa for analyzing the time series and helping me to interpret the results.

Let us start with the idea of a recurrence plot. Let's assume that we have a time series describing the appearance of six different types of discrete behavior. In our case, we measure the appearance of five different emotion types (Anger, Disgust, Fear, Joy, Sadness) and a neutral state N, where no clear emotion has been detected. The measurements are conducted by following a person's emotional state over time. Here is a small example of such a series:

N A N A A A D A F J J S S N N N

Now, we are interested to find whether we may identify a *pattern* of repetitions. Now let us use a sliding window (see Chap. 3) of size 1 and move it along our time series looking for places in which there is a recurrence of a specific emotion. Here is a rough animation of the first five steps, where the sliding window is marked using the strikethrough effect:

1. ~~N A~~ N A A A D A F J J S S N N N

2. N ~~A N~~ A A A D A F J J S S N N N

3. N A ~~N A~~ A A D A F J J S S N N N

4. N A N ~~A A~~ A D A F J J S S N N N

5. N A N A ~~A A~~ D A F J J S S N N N

What we see is that, at steps 4 and 5, anger repeats itself. We may represent the recurrence of values by using a two-dimensional plot where the X axis represents time point i and the Y axis represents time point j. In this case, we mark with a point each instance where we have a recurring value and the outcome of this representation is a "recurrence plot." In other words, a recurrence plot is a binary square matrix that represents all pairs of observations at time i and j that are either very similar—that is, recurrent ($R(i, j) = 1$ when $x(i) \approx x(j)$) – or dissimilar ($R(i, j) = 0$ when $x \neq 0$). Several measures are available that allow recurrence plots to be quantified and that are suitable to uncovering recurrence properties (Marwan et al. 2007). Figure 10.3 is just one example of a recurrence plot.

The very simple procedure described above can give rise to an extremely rich process of analysis in which various measures of recurrence are extracted from a time series. For instance, the "recurrence rate" is the fraction of recurrences found— that is, it is the density of recurrence points in the recurrence plot. It can be interpreted as the probability that the system will recur to an arbitrary state. The measure "determinism" is the fraction of recurrence points that form diagonal lines in a recurrence plot. Such diagonal lines are characteristic of processes of a predictable nature and are related to the divergence behavior of the system. The "mean line length" of such diagonal lines therefore reflects the average timescale of possible predictions. In addition to diagonal lines, vertical and horizontal lines in a

Fig. 10.3 An example of a
recurrence plot

recurrence plot can be used for quantification, because such lines are related to very
slow changes in the system. The measure "laminarity" (abbreviated LAM) is the
proportion of recurrence points that form such vertical lines and can be interpreted
as the probability that a randomly selected state of a system will not change for
some time. The vertical distances between recurrence points (=white vertical lines
on a black–white representation of the recurrence plot) are simply the recurrence
times of the states of the system. The distribution of recurrence times contains
additional information about the dynamics of the system. The "recurrence time
entropy" is the Shannon entropy of this distribution and is a measure of repeti-
tiveness. The last measure, "transitivity," is based on applying ideas from complex
networks to the recurrence analysis by identifying the recurrence plot with the
adjacency matrix of an undirected, unweighted complex network (Marwan et al.
2009). The transitivity coefficient measures is a network measure that provides the
probability that two neighbors of a node are also connected with themselves (i.e.
forming a triangle). From the viewpoint of recurrence plots, it gives the probability
that three close recurrence points will remain close (but only roughly speaking).
Systems with high regularity have a high transitivity coefficient, whereas irregular
dynamics will result in low values. This transitivity measure characterizes the
geometrical properties of the dynamic system in the phase space. These recurrence
quantification measures can be calculated using moving windows that slide over the
time series. Variations in these measures provide insights into the varying dynamics
of the underlying process and can be used to compare two different systems
regarding this aspect.

Now, do you remember the conversation between the pedophile and his victim?
My colleagues and I applied RQA[2] to these time series and to the expression of lust
in the conversation by each of the interlocutors. As we remember from the previous
irreversibility analysis, it seems that the pedophile mind is much more simple as he

---

[2]I'm grateful to my colleague Dr. Norbert Marwan for conducting the RQA analysis of these time
series and helping me to interpret the results.

Fig. 10.4 The evolution of
the laminarity measure in the
pedophile and the victim's
conversation
(LAM = Laminarity)

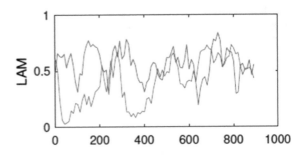

is "locked" on his prey and focused on his sexual desire. It is interesting to examine whether this conclusion can be supported by the RQA analysis. Figure 10.4 presents the evolution of the laminarity measure. In the figure, the red line designates the pedophile and the blue line his victim. Regardless of the extent to which each of the interlocutors in the conversation uses "dirty" language, we can see that they present different evolutions of the laminarity measure. The pedophile presents higher levels of laminarity at the beginning of the talk. As laminarity indicates a length of time in which a state doesn't change (or changes very slowly), we may infer that lust, at least as expressed by the pedophile at the beginning of the conversation, remains much higher than the lust of the victim, indicating that he is "locked" on the level in which he expresses his lust. As you can see, he presents a higher level of laminarity [mean = 8.63 (SD = 2.6) vs. 7.70 (SD = 2) respectively].

The pedophile also scored higher on the transitivity measure [0.56(0.13) vs. 0.40 (0.20) respectively]. To recall, systems with high regularity have a high transitivity coefficient whereas irregular dynamics will result in low values. The pedophile seem to be highly "regular" when his lust is expressed.

While characterizing dynamic patterns of personality may be highly important for personality analysis, we should recall that the dimensions of personality don't work as isolated parts but rather in concert, in a way we should better understand. The next section introduces the concept of synergy that may deepen our understanding of personality as a whole which is different from the sum of its parts.

10.3   Synergy and Personality: The Whole Is Different from the Sum of Its Parts

In this book and in other places (Neuman, in press), I've repeatedly criticized the simplistic theorization of personality and the way personality is measured. Indeed, anyone who has tried to seriously develop systems of computational personality that go beyond simple classification of tagged corpora knows that personality is much more complex than we usually conceive. But the fact that human personality is complex doesn't mean that we have to neglect any attempt to study it. In contrast, it is a starting point that should motivate us to seek better methodologies in order to address this challenge.

One of the most important concepts that may help us to pave the way is "synergy." Let me introduce this concept informally. When we study a "complex" object, we realize that its characteristics cannot be reduced to its components. Think, for instance, about a molecule of water, which is composed of one oxygen atom and two hydrogen atoms. The molecule is a composite structure that has unique properties (e.g., wetness) that cannot simply be reduced to the properties of its atoms or their sum. This means that, when we are dealing with complex objects such as personality, we should remember the old gestalt idea of a whole that is different from the sum of its components. This idea is probably valid for the study of personality, which is a complex object. One may wonder why is it important to deal with the complexity of personality when we have the full arsenal of ML and the simple task of classification. The answer is that there are many tasks that are too complex to address through these tools and methodologies. When we would like to screen in advance for individuals who might pose a danger to other people, we encounter a problem. It is clear that these individuals express a *unique mixture* of flammable ingredients. Each of these ingredients may have minor informative value when taken as an isolated feature but their combination, given a certain trigger, leads to a violent explosion. Interestingly, the study of personality in psychology has adopted a simplistic conception that doesn't deal with real-world complexity and ignores the synergetic effect of personality dimensions.

Synergy is a special case of irreducibility to atomic elements, but it is the collective *action* of the elements that is irreducible and not the elements themselves (Griffith 2014). How can we measure such a synergetic effect? There are various measures of synergy, but let us adopt the one by Williams and Beer (2010; elaborated by Timme et al. 2014). For a simple presentation of synergy and its measurement, let us limit ourselves to the case of two variables—X1 and X2—that supposedly provide information about a third variable, which is Y.

From the perspective of information theory, redundancy is the information received by Y from both X1 and X2 while synergy is the "bonus" information received from knowing how X1 and X2 work together (Timme et al. 2014). Think, for example, about the attempt to understand an obsessive personality through two behavioral indicators: cleaning and response to a violation of order. Let us assume that Greta lives on her own in a 50-square-meter apartment in Berlin. When analyzing her monthly expenses, we find out that her expenses on hand soaps, body soaps, dishwasher materials, and toilet- and bath-cleaning materials are higher by two and a half standard deviations than the population mean, while her other expenses are quite similar to the population mean. This behavioral pattern (X1) may be indicative of an obsessive–compulsive personality (Y). Now, let us assume that Greta is taking a part in a psychological experiment in which her heartbeat is measured before, during and after an experimental manipulation in which her well-organized bathroom is totally messed up by two enthusiastic teenagers. In this case, we should expect that the change in her heart beat (X2) will be a good indicator of her personality (Y). However, it is also reasonable to expect that both the behavioral pattern (X1) and the result of the psychological experiment (X2) will be redundant in the sense that our ability to identify Greta as an obsessive personality leans on information that exists in both of the two indicators. This redundancy may even motivate us in future applications to focus on shopping behavioral patterns rather than on the more expensive psychological experiment.

In contrast, let us imagine another situation, in which we are interested in predicting the psychopathic personality. We use two personality items where the individual is asked to rate his agreement with the following statements:

X1: I believe that taxidermy is a wonderful hobby.
X2: I consider cherries to be the most erotic fruit.

The individual's responses to these two items may share no information with regard to the psychopathic trait. However, the responses may have a synergetic effect and *together* may provide us with information about psychopathy.

Let us start by recalling the entropy measure that quantifies the uncertainty of a variable:

$$H(X) = -\sum p(x) \log(p(x))$$

where the entropy is maximized if the probability distribution is uniform. The next term to be introduced is mutual information, which is the amount of information provided about one variable when one knows another variable:

$$I(X;Y) \equiv H(X) + H(Y) - H(X,Y)$$

Williams and Beer (2010) suggest that the mutual information between variables X1, X2 and Y can be written as follows:

$$I(X1, X2; Y) \equiv Synergy(Y; X1, X2) + Unique(Y, X1) + Unique(Y, X2)$$
$$+ Redundant(Y; X1, X2)$$

They also suggest that redundancy is equal to a new information expression that is the minimum information function actually derived from the measure of specific information:

$$Ispe(y; X) = \sum_{x \in X} p(x|y) \left[\log\left(\frac{1}{p(y)}\right) - \log\left(\frac{1}{p(y|x)}\right) \right]$$

This measure indicates the amount of information provided by X about a specific state of Y. For example, if X is a variable indicating extraversion (1 = extravert, 0 = non-extravert) and y indicates a specific value of neuroticism (y = neurotic), then the above measure of specific information quantifies the amount of information provided by the extraversion variable about being neurotic.

Next, we introduce the measure of minimum information:

$$Imin(Y;X1,X2) \equiv \sum_{y \in Y} p(y) minxi\, I\, spec(y;X1)$$

This is the amount of information provided by the different X variables for each of the Y variables considered individually. As argued by Williams and Beer (2010), the minimum information is equal to redundancy:

$$Redundancy \equiv (Y;X1,X2) \equiv Imin(Y;X1,X2)$$

The minimum information is the average minimum amount of information about Y that can be obtained from any of the X variables. After we have calculated the redundancy measure, we can calculate the synergy and unique information measures as follows:

$$Synergy(Y;X1,X2) = I(Y,X1,X2) - I(Y,X1) \\ - I(Y,X2) + Redundancy(Y,X1,X2)$$

$$Unique(Y,X1) = I(Y,X1) - Redundancy(Y;X1,X2)$$

$$Unique(Y,X2) = I(Y,X2) - Redundancy(Y;X1,X2)$$

Applying this information measure to personality indicators, we may be able to determine which indicators provide us with unique information on personality (and how much); to calculate the redundant information that exists in our system; and, most importantly, to check whether any unique indicators operate synergistically to inform us about an individual's personality. I don't propose the above ideas as the only way of measuring synergy or even as the best way. The above measures are imbued with various difficulties. However, the approach presented by Williams and Beer (2010) at least attempts to formalize synergy in a structured way.

In sum, in this chapter, I've pointed out several novel ways in which the complexity of personality may be studied. These directions are in their infancy and much effort is needed to bring them to maturity and to prove their relevance to real-world challenges involving computational personality tasks.

References

Griffith, V. (2014). Quantifying synergistic information (Doctoral dissertation). California Institute of Technology.

Inches, G., & Crestani, F. (2012). Overview of the international sexual predator identification competition at PAN-2012. In P. Forner, J. Karlgren & C. Womser-Hacker (Eds.), CLEF (Online Working Notes/Labs/Workshop) (Vol. 30). Rome, Italy.

Lacasa, L., Nunez, A., Roldán, É., Parrondo, J. M., & Luque, B. (2012). Time series irreversibility: A visibility graph approach. *European Physical Journal B, 85*(6), 1–11.

Marwan, N., Donges, J., Zou, Y., Donner, R., & Kurths, J. (2009). Complex network approach for recurrence analysis of time series. *Physics Letters A, 373*(46), 4246–4254.

Marwan, N., Romano, M. C., Thiel, M., & Kurths, J. (2007). Recurrence plots for the analysis of complex systems. *Physics Reports, 438*(5/6), 237–329.

Nietzsche, F. (1882/2010). The gay science: With a prelude in rhymes and an appendix of songs. New York, NY: Vintage.

Neuman, Y. (in press). Shakespeare for the intelligence agent. Lanham, MD: Rowman & Littlefield.

Timme, N., Alford, W., Flecker, B., & Beggs, J. M. (2014). Synergy, redundancy, and multivariate information measures: An experimentalist's perspective. *Journal of Computational Neuroscience, 36*(2), 119–140.

Webber Jr., C., & Zbilut, J. (2005). Recurrence quantification analysis of nonlinear dynamical systems. In M. Riley & G. Van Orden (Eds.), Methods for the behavioral sciences (pp. 27–94). Eds. Washington, DC: National Science Foundation.

Weiss, G. (1975). Time-reversibility of linear stochastic processes. *Journal of Applied Probability, 12*(4), 831–836.

Williams, P. L., & Beer, R. D. (2010). Nonnegative decomposition of multivariate information. arXiv preprint arXiv:1004.2515.

Chapter 11
Discussion

Personality is our way to conceptualize relatively stable patterns of thought, emotion and behavior. From a philosophical perspective, personality is a descriptive, explanatory and predictive concept. First, it is used to make sense of others' and the self's behavior by mapping fragments of thoughts, emotions and behavior into a coherent pattern, a *narrative* of self and others. When I describe someone as a "narcissist," I'm actually trying to describe this person through a coherent narrative summarized through a single personality tag. However, personality isn't just a descriptive tool used to make sense of others. It is also an explanatory tool through which we may explain the *reasons* we choose to behave as such and such. For instance, in an attempt to explain the motivation of a certain individual to be a poet, we may point to his "introverted" personality and his deep interest in self-reflection. The third justification, and probably the most important one, concerns the use of the concept "personality" to *predict* future moves—and, to add another layer, to *control* it. Prediction and control go hand in hand. We would like to predict the future in order to remove some of the uncertainty associated with it, and to control our and others' destiny. A talented clinical psychologist, such as the one played by Robin Williams in the beautiful movie *Good Will Hunting* (Bender and Van Sant 1997), tries to understand his patient's personality in order to control him—that is, to change him for good purposes. In this context, the field of automatic personality analysis—or what we might call "computational personality analysis"—is located at an interesting historical junction where the importance of understanding individuals is converging with the emergence of rich datasets and powerful data-analysis algorithms.

In the introduction, I argued that we now ask the question "Who are you?" rather than "What are you?" and that this changing perspective provides us with a variety of challenges and opportunities, as is evident in personalized medicine and hopefully in computational psychology. It is important to remember that, when dealing with psychology in general and personality analysis in particular, the individual is our basic unit of analysis, and the development of tools of computational personality should respect this scale of analysis. In the era of Big Data, computational

© Springer International Publishing Switzerland 2016
Y. Neuman, *Computational Personality Analysis*,
DOI 10.1007/978-3-319-42460-6_11

tools involve the processing of large text collections and may not be adequate to the analysis of human personality. This point can be illustrated via analogy to automatic text comprehension.

As argued by Berant et al. (2014), work in the field of text comprehension has usually been focused on "macro-reading," while "micro-reading," involving deep understanding of a single text, is currently beyond the scope of state-of-the-art text-processing systems. Is the difference between the micro and the macro levels of analysis a semantic issue per se? Is it an issue relevant to computational personality analysis?

Let us assume that you are invited by MI5 to advise its employees about computational tools to analyze threatening letters sent to the Queen of England. The situation is simple. Each year the Queen probably receives hundreds of letters threatening her life. Many of these letters are not anonymous and many of them threaten the Queen as their writers believe that they are the true heirs to the crown. Now, in dealing with these letters, one has to perform a risk analysis in order to set priorities, as the authorities cannot seriously handle the potential threats of each and every disturbed person who has written a letter. The basic ML approach is to use a tagged corpus of individuals who actually tried to act violently against the Queen in the past and to train the classifier on the letters sent by these individuals. New letters can then be analyzed automatically in order to determine how likely they are to belong to the category of potentially dangerous individuals.

However, things may change, and into the "game" may enter radical Islamists who threaten the Queen not because they believe that the crown has been stolen from them but because they believe that the Queen represents Western imperialism, which they consider to be an evil force. In this context, we may train the classifier on a new corpus. However, and luckily, up to now, no terrorist attack has been launched against the royal family by Islamic terrorists who sent a letter in advance, and so we have no corpus for training our classifier. A pragmatic solution is to include in the threat-analysis system a micro-reading module that reads and processes each letter to better understanding its meaning, its underlying psychological structure and the threat it might possess. For example, let's take a letter of threat sent by jihadists to the spouses of British servicemen in July 2015:

1. You have been identified as an unbeliever and a bride of a murderer of the servants of Allah in the Holy Land.
2. Our peoples have suffered at the hands of your husbands who have murdered, killed and raped our women and children.
3. We the servants of Allah intend to avenge our peoples by destroying the families of unbelievers in the land of hate.
4. You along with many others will pay the price for your husband's destruction in the Holy Land.
5. We now know where you live and will begin to destroy the unbelievers and their families as they have done to us.

Table 11.1 Analysis of the threat letter

Who	You and your husbands
	We the servants of Allah, our wives and children
When and what	In the past you murdered, killed and raped and we suffered
	In the future, we intend to avenge and you will pay the price
Why	Revenge
Where	The holy land
	The land of hate
	Where you live
How	Not specified

From the first sentence we may extract the following propositional knowledge:

unbeliever[You]
are, of_a[You, bride, murderer]
of[murderer, servant of Allah in the Holy Land]

The second sentence can be represented as follows:

suffered [our people, husbands]
murdered, killed, raped [husbands, our women and children]

At this point, we may identify three main actors:

1. "you,"
2. the murders/husbands and
3. the servants of Allah, their wives and children, approached from the perspective of "we."

In terms of the use of past tense, the "you" has been "identified," the servants and their families have suffered and the husbands have murdered, killed and raped. Moving to the future tense, we find that the "servants" intend to "avenge" and "destroy" and that "you" will pay the price. From this micro-reading of the threat letter, we may clearly elucidate violent intentions motivated by revenge. Asking the "who" question may provide us with three main actors; asking the "why" question may provide us with the motivation, which is revenge, as shown in Table 11.1.

We can see that a micro-reading of the text provides us with the ability to better understand the threat. Moreover, while the letter clearly expresses the intention of getting even, the "how" question remains unanswered. This is an important point in terms of scoring the level of threat and how serious the intentions are. Details and realization of the act's consequences are two important dimensions to trying to understand how serious a threat is.

11.1   From What to Who and from Why to How

I have pointed to the shift from the "what?" to the "who?" question, but I would now like to propose another shift: from the "why?" to the "how?" question. To explain this point, let me draw on an important book about an extensive subject of study among historians: World War I. In his bestseller *The Sleepwalkers: How Europe Went to War in 1914*, Clark (2013, xxvii) writes:

> Questions of why and how are logically inseparable, but they lead us to different directions. The question of *how* invites us to look closely at the sequences of interactions that produced certain outcomes. By contrast, the question of *why* invites us to go in search of remote and categorical causes.

Clark's observation is highly relevant to the analysis of human behavior in the context of computational personality. Asking why a certain individual performed a certain atrocity, such as murder, is probably an inevitable question for intelligent creatures who seek to identify causes. But what is the psychological cause of such an atrocity? Genetics? Pathological parenting style? A childhood trauma? A combination of these causes? Or can we merge all of these causes under the tile of a "sociopathic personality"?

The term "sociopathic personality" is an important heuristic, and it allows us to move from seeking among numerous potential causes of behavior to a concise and informative concept, which is sociopathy/psychopathy. However, what Clark is saying is that, instead of focusing on "categorical causes," we may gain deeper insights if we seek the "sequences of interactions that produced certain outcomes." This idea, proposed by a historian, is strikingly similar to the idea of a "reactive system" in biology (e.g. Cohen and Harel 2007), which has been successfully used in modeling biological systems.

What does it mean to identify "sequences of interactions" and their outcomes in the context of personality? Let me explain this possible direction through a concrete example. Let's assume that we are seeking to identify depression among teenagers before it reaches a clinical tipping point. Let us also assume that we have all the data we need, including medical records, school records, communication patterns in social media, Instagram data and so on. In this case, we may be interested in identifying temporal patterns of interactions between variables leading to certain outcomes of interest.[1] For example, we may find that reported absenteeism from school interacts with certain music types to which the subject is listening, and there may be certain textual features in the subject's messaging that predict outbursts of clinical depression. In this case, we may use hundreds and thousands or even hundreds of thousands of features in order to identify temporal *patterns of interaction* associated with certain outcomes.

The features issue is crucial for this task. Currently, models of computational personality are "flat" in the sense that abstraction has no place in the design of a model. As argued in the context of "executable biology" by Fisher et al. (2011, p. 3), "Abstraction is well understood to be a key to modeling complex systems… where by 'abstraction' we refer to a model at a certain level of description, suppressing lower-level details in a principled way." Given this insight, our future models should have a strong element of intelligent abstraction that may allow us to

---

[1]For a similar approach albeit in a different context, see Figueiredo et al. (2015).

model the multi-scale structure of human behavior.[2] Given features and their abstracted representations, we may seek "sequences of interactions." Time, interaction and the unique information synergy identified through these temporal interactions may together be the key to future technologies.

11.2   Ethical Considerations

In this brief manuscript, I've attempted to introduce the field of computational personality analysis from a broad theoretical and critical perspective and by presenting several case studies and my knowledge and experience. It is clear that this field is taking its first steps and that it should make further steps in order to study human personality as a complex phenomenon through computational tools. Let me conclude by touching on the ethical issue.

When giving a talk about computational personality analysis to an audience of university faculty members and students, I have encountered the question of how it is possible to study human personality if we assume that humans have free will and the ability to determine their course of action. I see no conflict whatsoever between the philosophical assumption of free will and a pragmatic interest in computing human personality. Moreover, the computational analysis of personality applies stochastic procedures to the rational and freely willed behavior of human beings. This move may be fully justified and the reason may be provided from a much more general perspective, which asks how is it possible to study a deterministic process through a stochastic perspective. As nicely explained by Yan et al. (2016), individual behavior may be entirely deterministic, but in each case "large numbers of possible choices exist which leads to a huge number of combinations. So when considering a large collection of realization of these choices, we cannot distinguish between underlying proper stochastic processes and deterministic processes." The point is that each individual may determine his choices through the mechanism of free will. However, our analysis doesn't concern a single individuals but collections of individuals, whose behavior, beliefs and emotions have some form of distribution that can be studied and used in a way that is indifferent to the underlying dynamics that produced them.

Another question that I'm regularly asked concerns the dangers of using technologies of computational personality—specifically, the possibility of them falling into the hands of tyrant regimes. The answer is simple, although the ethical burden accompanying it doesn't become lighter. Human technology has always had a Janus face; fire can be used to warm your body and to cook food but it can also be used to

---

[2]This process of abstraction may be modeled using a variety of mathematical tools such as the hypergraph, which is a generalization of a graph in which an edge can connect several vertices. If an edge represents certain semantic relations between entities, we may group and abstract vertices connected by the same edge.

burn down a house; an ax can be used both to cut wood and to murder; and a computer can be used to develop new drugs against cancer or as a tool of communication between vicious terrorists. Trying to block the advancement of technology is a naive move that is blind to the infinite desire of human beings to gain power. It is better to try to seek ways in which these technologies will be constrained, regulated and "sublimated" (to use a psychological term) for the benefit of human beings, assuming that misuse and abuse are *inevitable* and that technological environments should be designed, monitored and regulated in such a way that minimizes risk and wrongdoing as much as possible.

In Judaism, the Midrash is a body of exegesis of the Bible along with homiletic stories. One of these texts is Kohelet Rabbah (or Ecclesiastes Rabbah), which dates to approximately 500–1000 CE. This text includes the following story (my translation):

> When the one (blessed be he) created the first man, he took him and returned him to see the trees of the Garden of Eden, and told him: See my deeds how pleasant and excellent they are, and everything that I've created. For you I've created. Be careful not to damage and destroy my world, as if you damage it there will be no one to fix it afterward.

This is a striking moral lesson as it was written hundreds of years before modern civilization—with its science and technology—started to threaten its own existence. Here, the Midrash provides us with foresight and a moral lesson in a period when the modern developments of human technology and their impact on the world could not even have been imagined. This lesson concludes this book, offering a moral imperative to pay careful attention to the technologies we develop and to make sure they are used for construction and good rather than for destruction and wrongdoing.

References

Bender, L. (Producer) & Van Sant, G. (Director). (1997). *Good will hunting*. USA: Lawrence Bender Productions.

Berant, J. et al. (2014). Modeling biological processes for reading comprehension. In *Proceedings of the 2014 Conference on Empirical Methods in Natural Language Processing, Doha, Qatar, 25 to 29 October 2014* (pp. 1499–1510). Stroudsburg, PA: Association for Computational Linguistics.

Clark, C. (2013). *The sleepwalkers: How Europe went to war in 1914*. London, UK: Penguin.

Cohen, I. R. & Harel, D. (2007). Explaining a complex living system: Dynamics, multi-scaling and emergence. *Journal of the Royal Society Interface, 4*(13), 175–182.

Figueiredo, F., Ribeiro, B., Almeida, J. & Faloutsos, C. (2015). TribeFlow: Mining & predicting user trajectories. *arXiv preprint arXiv:1511.01032*.

Fisher, J., Piterman, N. & Vardi, M. Y. (2011). The only way is up. In M. Butler & W. Schulte (Eds.), *FM 2011: Formal methods* (pp. 3–11). Berlin, Germany: Springer.

Yan, X., Minnhagen, P. & Jensen, H. J. (2016). The likely determines the unlikely. *arXiv preprint arXiv:1602.05272*.

Appendix
Snowden's Letter

To whom it may concern,

I have been invited to write to you regarding your investigation of mass surveillance.

I am Edward Joseph Snowden, formerly employed through contracts or direct hire as a technical expert for the United States National Security Agency, Central Intelligence Agency, and Defense Intelligence Agency.

In the course of my service to these organizations, I believe I witnessed systemic violations of law by my government that created a moral duty to act. As a result of reporting these concerns, I have faced a severe and sustained campaign of persecution that forced me from my family and home. I am currently living in exile under a grant of temporary asylum in the Russian Federation in accordance with international law.

I am heartened by the response to my act of political expression, in both the United States and beyond. Citizens around the world as well as high officials—including in the United States—have judged the revelation of an unaccountable system of pervasive surveillance to be a public service. These spying revelations have resulted in the proposal of many new laws and policies to address formerly concealed abuses of the public trust. The benefits to society of this growing knowledge are becoming increasingly clear at the same time claimed risks are being shown to have been mitigated.

Though the outcome of my efforts has been demonstrably positive, my government continues to treat dissent as defection, and seeks to criminalize political speech with felony charges that provide no defense. However, speaking the truth is not a crime. I am confident that with the support of the international community, the government of the United States will abandon this harmful behavior. I hope that when the difficulties of this humanitarian situation have been resolved, I will be able to cooperate in the responsible finding of fact regarding reports in the media, particularly in regard to the truth and authenticity of documents, as appropriate and in accordance with the law.

© Springer International Publishing Switzerland 2016
Y. Neuman, *Computational Personality Analysis*,
DOI 10.1007/978-3-319-42460-6

I look forward to speaking with you in your country when the situation is resolved, and thank you for your efforts in upholding the international laws that protect us all.

With my best regards,

Edward Snowden, 31 October 2013

Author Index

© Springer International Publishing Switzerland 2016
Y. Neuman, *Computational Personality Analysis*,
DOI 10.1007/978-3-319-42460-6

Subject Index

A
Analogy, 72–77, 110
Artificial intelligence, 6, 92
Attractors, 79, 85, 87–92, 95, 97
Augmented intelligence, 92
Authorship attribution, 69

B
Big data, 61, 109
Bin Laden, 18, 88, 91, 92

C
Cho, 91
Cognitive–behavioral approach, 20, 21, 23
Columbine high school massacre, 50
Conversion, 93, 97
Curse of dimensionality, 30, 31

D
Deep learning, 31, 32
Defense mechanisms, 16
  humor, 16, 17
  projection, 18, 91
  splitting, 90
Dependency parsing, 64, 66
Deterministic tourist walk, 87–89, 94
Dissipative system, 79
Distributional semantics, 43, 51, 52, 58

E
Emotional valence, 11, 90
Entropy, 35, 36
  permutation entropy, 34–36

F
Factor analysis, 12
Features selection, 30, 31
Five Factor Model of Personality
    (Big Five), 5, 12, 23
  agreeableness, 12
  conscientiousness, 13
  extraversion, 12, 14
  neuroticism, 12–14
  openness to experience, 12, 79

I
Intention, 18, 46, 49–51,
    82, 111
Irreversibility, 79, 99–101, 103

L
Law of requisite variety, 23
Lexical approach, 12
LIWC, 30, 32, 95

M
Machine learning, 6, 27
Mental models, 9, 10, 12, 23
Metaphor, 72, 74
Music, 43–45, 112

N
Narrative, 17, 47, 81–84, 90
Natural language processing,
    6, 69, 81
Network motifs, 54, 55, 58
n-grams, 24, 32, 69

© Springer International Publishing Switzerland 2016
Y. Neuman, *Computational Personality Analysis*,
DOI 10.1007/978-3-319-42460-6

Printed in the United States
By Bookmasters